THE FORTH AND CLYDE CANAL

THE FORTH AND CLYDE CANAL
A HISTORY

T. J. DOWDS

TUCKWELL PRESS

First published in Great Britain in 2003 by
Tuckwell Press Ltd
The Mill House
Phantassie
East Linton
East Lothian, Scotland

Copyright © T. J. Dowds 2003

ISBN 1 86232 232 5

British Library Cataloguing-in-Publication Data

A catalogue record of this book is available
on request from the British Library

Typeset by Hewer Text Ltd, Edinburgh
Printed and bound by Bell & Bain Ltd, Glasgow

Contents

Illustrations

Acknowledgements

There are many people without whose help this book would never have seen the light of day, and whose contribution must be recognised. Firstly, I am indebted to Professor Roy Campbell and Dr Anand Chitnis, who were then on the staff of the University of Stirling and who supervised, guided and criticised the research that forms the basis of the present work. Like most historians, I am greatly indebted to the members of staff of several libraries and institutions, who gave so willingly of their time and expertise to assist with the research: viz. the Library at the University of Stirling, the Andersonian Library, University of Strathclyde, Glasgow University Library, The Mitchell Library, Glasgow, The National Library of Scotland, Edinburgh, and the William Patrick Library, Kirkintilloch. The unfailing cooperation of the staff at Falkirk Museum, before and after its move to Callendar House, the Scottish Record Office (now the National Archives of Scotland), and the House of Lords Record Office, London, was much appreciated. My debt to each and every one of them is enormous, and I hope that the end product justifies their efforts, although any errors and omissions are the fault of the writer, not of those who supplied the information. Finally, but by no means least, I must thank my family, particularly my wife, Cathy, who has patiently tolerated the hours spent in libraries and museums, the trips to sites along the Canal, and the pounding of the keyboard at all hours, as the work was brought to completion.

Introduction

At the start of the new millennium, there was an increasing interest in Scotland's inland waterways, highlighted by the plan to reopen part of the Forth and Clyde Canal for leisure purposes and by the construction of a spectacular 'wheel' at Lock 16, its junction with the Union Canal. The consequent publicity resulted in a greater awareness of the canal and a growing interest in its, often neglected, history.

For just under a century, canals, or inland navigations as they were more commonly then called, provided the only means of transporting heavy goods from one part of the country to another. The Forth and Clyde Canal allowed coal and machinery from Lanarkshire and the western industrial centres to be carried to Edinburgh and the east, and the grain from the richer farming areas in the Lothians to reach the densely populated area around Glasgow, while at the same time allowing passengers to travel between Glasgow and Edinburgh in greater comfort than was afforded by the stagecoach, which took almost 14 bumpy hours to complete the journey. It also brought employment opportunities along its banks with the need to provide lock-keepers, stable-hands, bargees, warehousemen and labourers to maintain the waterway and its traffic. Consequently, towns and villages came to rely on the canal for a large part of their prosperity. But this came to an abrupt end with the onset of 'railway mania' in the nineteenth century.

The arrival of the steam engine and its associated network of lines that connected almost every town and village in the country made the Canal obsolete. Railways were cheaper and faster than travel by canal and freight and passengers took to the new mode of travel. For over a century afterwards, the canals lay idle, ignored by business and neglected by the public. The only time that they received publicity was when children, attracted to the water to play or to fish, fell in, sometimes with fatal results, and this prompted demands for their closure. By the mid-1950s, residents in some of the new housing schemes that had been built along the banks of the canals regarded them as dangerous and sought to have them filled in, while others saw their conversion to motorways as a means of improving road traffic at less cost than was normally associated with road construction. But for most of the time, the public was only vaguely aware of the existence of these formerly important links.

At the same time, there was a growing awareness among historians that Scotland's economic history had not been given the prominence that it deserved, and that the re-development that was then taking place was often destroying all trace of the industrial past. Consequently, attention was focused on researching the industrial history of the country and, where necessary, recording the structures before they disappeared – what might be termed 'rescue industrial archaeology'. One of the obvious areas that attracted attention was the communication system provided by the canals, and their role in making possible the industrial revolution in Scotland.

Two distinct groups of historians tackled the problem. First, those whose main concern was the preservation, or at least recording, of the threatened structures and their associated workings. They collected photographs, memoirs from the towns and villages along the route in an effort to preserve the memory of a bygone age. Others looked at the place of the Canal in the formation of industrial Scotland in an effort to understand and explain how the Canal came into being, and its role in the life of the country. To this latter group of historians, the Forth and Clyde Canal was seen as essentially a means of linking the east and west coasts, with the Union Canal providing a route to the capital and the Monkland Canal opening up the industrial area of Lanarkshire.

This book sets out to answer three basic questions: why was the Canal built? (An examination of the motives of those who proposed the scheme.) Secondly, why was it built in the 1760s? (Why had it taken a century from the first proposals until work commenced?) Thirdly, why did it take 22 years to build?

To reach an answer to these questions, it was necessary to look at the state of the Scottish economy at the start of the eighteenth century to see if a scheme of this magnitude was feasible. It became clear that such an undertaking was not possible before there were sufficient funds to finance the project and, perhaps as important, a will to undertake the task. The backwardness of the largely agricultural economy was an obstacle not overcome until the commercial benefits of the Union of 1707 had produced the wealth that allowed funds to become available. By the middle of the century, there had arisen a class of merchants, mostly in Glasgow, with the surplus money to finance the venture and who were prepared to wait for some years before they would see any return on their investment.

Once there was a proposal, other factors came into play. A bitter row arose between Scotland's two leading cities over the scheme, with Glasgow seeing the Canal chiefly from its own commercial standpoint, and Edinburgh viewing it as a matter of civic pride. This debate could be just as easily interpreted as a power struggle between them to determine which of the cities was to play the leading role in the project: the old capital or the commercial newcomer. The Government had a different perspective. It was required to pass the legislation required to build the Canal and saw in the scheme a means of opening up Scotland to more 'British' influences and so ending any lingering attachment to the House of Stuart – the Jacobite Rising had taken place only 22 years earlier and there was still a nostalgia, even sentimentality, about 'Bonnie Prince Charlie'. With the need to involve Westminster, London became the third city to have an active role in the making of the Forth and Clyde Canal.

Once construction began, other problems emerged. The route presented more difficulties than had been anticipated, not least the geology of the countryside through which it had to pass. Dullatur Bog presented drainage problems that ate up much more time and money than was envisaged by the surveyors. At a number of places the channel had to be blasted through solid rock; again this had not been foreseen, and contributed to both expense and delay. Archaeology also provided reasons for delay, at least in the early stages. The discovery of bodies at Dullatur and Roman artefacts at a number of locations meant that work stopped until the finds had been recorded and removed. In addition, the labour force created a number of difficulties: largely inexperienced at first, the workers gained experience on the job, but at times disappeared to take on more congenial work on the land. Given the internal disputes between Edinburgh, Glasgow and London, as well as with some of the subscribers and contractors, a speedy completion was not likely.

Finance, however, was to be the greatest factor in delaying progress. The cost of the work had been seriously underestimated and, even when the subscriptions that had been promised were paid, which did not always happen, funds were never sufficient to complete the task. As a result, the Canal was completed in stages, and it was not until the Government decided to make money available, with the consequence that control passed to London, that the East and West Seas were finally joined in 1790.

The Forth and Clyde Canal was not Scotland's first inland waterway; that honour should go to the quarter-mile cut between Largo House and Church, which was opened in 1495 to allow the retired Admiral, Sir Andrew Wood, to sail his barge to church on Sundays. But the Forth and Clyde was by far the largest undertaking that had ever been seen in the country and as such presented problems that had never been encountered before. The need to keep three centres – Edinburgh, Glasgow and London – informed of progress led to the invention of novel methods of administration. The supervision of a huge workforce, not just the navvies, but also those constructing the ancillary facilities like wharves, stables and warehouses, forced the Canal Company to experiment with its organisation and produce a system never seen before. Such was its success that the Canal Company's administration became the model for all future large-scale undertakings. It is no exaggeration to say that modern business organisation began with the making of the Forth and Clyde Canal.

The Forth and Clyde Canal: An Overview

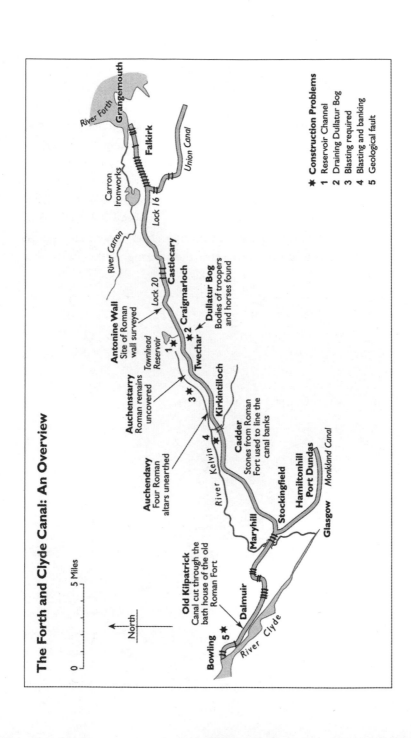

Scotland in the Early Eighteenth Century

Historians have described Scotland in the eighteenth century as being characterised by 'a gradually mounting tempo of Improvement'. The early years of the century witnessed a number of changes in rural society that became more widespread and common. Roads, bridges and harbours were constructed, canals were cut, and agricultural methods were reformed. It was this series of changes that was later referred to as the 'Improvement'. Contemporaries were, perhaps, less aware of this new spirit that historians have identified as influencing Scotland, and saw the changes as a response to alterations in the economic life of the community, rather than a conscious attempt to re-structure society. This process of change was not new, and had been evident in the previous century, as early agrarian changes in the late seventeenth century indicate. The ill-fated Company of Scotland's attempt to establish a trading colony on the Isthmus of Darien illustrates the increasing importance of economic factors in society, and of the increasing scale of such ventures that was to be characteristic of the 'Age of Improvement'.

Encouraged by 'improving lairds', like Cockburn of Ormiston, the Earl of Hopetoun and the Earl of Stair, there was a growing interest in agriculture. This interest was motivated partly by a desire for higher incomes from rents on the part of landowners and partly by a more general anxiety that there be no return of the famines of the 1690s. As the landed estate was the basic unit of the Scottish economy and the source of wealth and status, there was an urge to find a means of increasing productivity and so provide stability to an uncertain rural economy. Aware that new methods had been successful in England and Holland, a new generation of 'improving' lairds, motivated by economic aims, brought about an increasing number of changes over a wide area of the country, and these were to prepare Scotland for the full-scale Agrarian Revolution of the middle years of the century. This period has been called by Professor Smout 'the prelude to the take-off.'[1]

Improvement was undertaken as a deliberate policy by many landowners, but the depressed price of agricultural produce before 1750 made it a risky business. Cockburn fell victim to his own enthusiasm and was obliged to sell his estate in 1747 to meet the debts he incurred as a consequence of his innovations. It has been suggested that up to the middle of the century, improving was only a rich man's, non-paying hobby. But this is not a complete explanation of why there was a movement that stressed economic matters in Scotland at the time. As early as the last decade of the seventeenth century, Fletcher of Saltoun had urged that a greater concern for economic matters was essential in order to preserve Scotland, prevent emigration of her people and reduce her dependence on England. This concern was shared by a number of other writers contemporary with Fletcher, who put forward ideas that were later regarded as part of the 'Enlightenment'. The culmination of this movement toward a greater awareness of the vital role of the economy was the Treaty of Union in 1707, when the demand for access to English markets, especially in the colonies, overcame the reluctance of those members of the Scots Parliament who wished to see the two countries retain their separate identities through indepen-dence or a federal union, and resulted in the acceptance of an incorporating union and the loss of a distinct Scottish identity.

The Union of 1707 added momentum to the trend toward improvement. Previously, Scotland's chief exports had been black cattle and coarse linen, with England being the main market; indeed, the threat implied to this trade by the Aliens Act of 1705 was one of the decisive factors encouraging the Scots

Parliament to accept the Treaty of Union. Now that the Scots and English shared a common Parliament, the Navigation Acts that before had hampered Scottish trade were to be beneficial, and Scots merchants had access to English markets at home and overseas. As a result, the merchant class everywhere began to grow prosperous, and this was especially so in the case of Glasgow, where the Tobacco Barons gathered wealth from the trade with Virginia. Glasgow had begun its rise to pre-eminence during the seventeenth century, and by 1697 accounted for 40% of the taxes raised from the Scottish burghs. After 1707, the commercial classes assumed an ever more prominent role in the life of the country and came to possess the knowledge and skills required for further advances, and this class provided, in the words of T.M. Devine, the 'social and economic environment which was not resistant to change, whether in the cultural or economic spheres'.

Economic factors were evident in Edinburgh also, where men of rank and property displayed a new interest in civic leadership in the 1720s, when the Union was establishing political and economic stability. Interest in agrarian reform was encouraged by the Honourable Society of Improvers in the Knowledge of Agriculture, founded in 1720 to provide members with information on the newest, scientific methods of farming. Within a few years, however, the Society had extended the range of its interests to include the education of tenants, the organisation of the linen industry, and deep-sea fishing.

It has been suggested that the Society became a 'para-parliament' into which the aristocratic governing class diverted their energies, making economic improvement an alternative to politics, which was now closed to most of them as a result of the loss of the Edinburgh Parliament and the reduced numbers representing Scotland at Westminster. However, interest in these ventures was not new, but had been evident in the previous century; it merely became more obvious after the Union. Also, the main burden of the work of this and similar societies fell not upon the aristocracy, but upon the lesser gentry, like William Mackintosh of Borlum, Thomas Hope of Rankeillor and Robert Maxwell of Arkland, indicating that many of the nobility did not see such activity as a substitute for a political career. A number of societies were formed in imitation of this one, and by 1727, the Board of Trustees for Manufactures and Fisheries was giving financial aid to economic enterprises, although its limited funds restricted the endeavours of the progressive landowners who made up the membership of the Board.[2]

The Select Society was founded in 1754 to replace the Honourable Society which had ceased to function after 1745 partly as a result of the agricultural depression of 1740–42, and the Jacobite Rising that had threatened the political and economic security of both kingdoms. The founders, David Hume, the philosopher and historian, Lord Kames, High Court Judge and enthusiastic improver, and Allan Ramsay, the painter, had the aim of 'improving' members by reasoning and debate in order to 'discover the most effectual method of promoting the good of the country'. It attracted the nobility and the clergy, but above all, a number of lawyers, and debates moved from purely intellectual and cultural subjects to matters of economics, commerce and the prosperity of the country as a whole. Professor Skinner has shown that the literati, centred on the capital, were able to transform what he calls the 'civic humanism' of the late seventeenth and early eighteenth centuries into a philosophy that was adapted to the modern world. Civic humanism saw self-fulfilment as being achieved where the individual acted as a citizen and participated in the life of the community: this has been advanced as the reason for the political involvement of the aristocracy. In post-Union Scotland, the road of political involvement was closed and the literati sought to discover an alternative outlet to this activity. Historians like William Robertson and John Millar in Glasgow, together with David

Hume and Lord Kames in Edinburgh, stressed the importance of economic and cultural variables in generating social change in the past. They proposed that involvement with commerce and industry, far from being beneath the educated aristocrat, could be regarded as a means of fulfilling the civic role of the humanist ideal in modern society.

Adam Smith was to take this argument further, and advanced the notion of a commercial stage of civilisation. He suggested that participation in economic life helped produce independence and individual happiness, as well as preserving from decline a 'noble form of civilisation'. Even after the collapse of the Select Society in 1764, the idea of virtuous citizenship lingered on in the many groups, largely social, but with an interest in all matters affecting life, that sprang up to help society along the path of improvement and happiness. It is arguable that the nobility were conscious of a need to divert their energies from politics after 1707, as they had always taken a keen interest in the life of the local community and saw concern for every aspect of that life as part of the duty that fell on the landowner.

It is true that the work of these societies helped transform the poverty-stricken, backward, land-based economy of Scotland and prepared it for the introduction and maintenance of an industrial society. The agencies of this transformation were the Board of Trustees for Manufactures and Fisheries, the Highland Society, this latter being particularly concerned to eradicate the last vestiges of Jacobitism in the North, and the Board of Commissioners for the Forfeited Annexed Estates, which used the income from lands confiscated from their Jacobite owners to finance a range of improvements.[3]

Agricultural improvement progressed as English methods were popularised by civic-minded landowners like Grant of Monymusk and Barclay of Urie. Land was enclosed to make compact farms in place of the old open-field, shared system, and the new methods were quickly implemented as any benefits

went to the landowner. Marshes were drained to bring previously unproductive land under the plough, and a new crop
rotation gave still higher yields and at the same time provided
fodder for cattle during winter. The use of lime and manure
further increased the fertility of the soil and led to even greater
productivity which enhanced the incomes of the improving
lairds. From the middle of the eighteenth century funds became
available to finance agrarian changes, and many tenants were
persuaded by the profitability of the new methods to undertake
improvements for themselves. The landowner provided the
fixed capital, and by insisting upon a return for his investment
in the form of higher rent, forced the tenant to make best use of
his land.[4]

Changes on the land encouraged a range of other diverse
enterprises like mining and quarrying to exploit the natural
resources of the earth. Together with the reorganisation of the
home-based textile industry, this created a need for better
communications. The need to move goods in bulk more easily
and cheaply prompted a further spate of improvements. Farmers
required to have lime, manure and building materials brought to
their lands, and to have crops carried to market. Cloth merchants,
who had supplied the domestic worker with flax and collected
the finished linen on periodic visits, were unable to cope with the
demands of the factories that required more goods than could be
carried by a packhorse, and they sought facilities to carry bulk
goods. Where, previously, the drove road had been sufficient for
the needs of the main item of export, Highland Black Cattle, at
the same time providing convenient routes for the cadgers to
travel from place to place to meet the very limited demands of
inter-regional trade, they proved totally inadequate for the needs
of the new economy that was being shaped during the second
quarter of the century

Military requirements, however, prompted the first serious
consideration of communications in Scotland. In 1746, Lord

Advocate Grant persuaded the government to divert money from the forfeited estates of exiled Jacobites into schemes to build roads that would open the Highlands to southern influences in order to pacify that region that had been loyal to the House of Stuart and which had provided the armed force that had almost overthrown the Hanoverian regime. As a consequence, troops resumed the road-building programme that General Wade had begun in the 1720s but which had been allowed to lapse under the laissez-faire influence of Walpole. This work continued for over thirty years, and James Boswell and Dr Johnson noted, in the *Journal of a Tour Through the Highlands* in 1773, that they had passed a number of soldiers engaged in road-making near Glenmoriston.

Landowners were quick to realise that benefits could arise from improved communications, understanding that a prerequisite of industrial growth is adequate transport facilities. Improving lairds had shown an early awareness of the need to create an infrastructure to support economic developments, and from the seventeenth century there had been concern about constructing harbours on the Forth to facilitate the export of coal. The Dundas family, which was to figure prominently in the affairs of the Forth and Clyde Canal Company, had played a leading role, having Grangemouth built as a coal port. One of the features of the improvements was that leading families, like the Dundases, became involved in a multitude of schemes designed to benefit the economic life of the community as a whole, not merely themselves.

Support for the landed interest came from the merchant class in towns where the advantages of the easier carriage of goods were more obviously appreciated. Glasgow in particular had benefited from the opening of the English colonial trade to Scots as a result of the Treaty of Union, and by the middle of the eighteenth century a highly profitable trade in tobacco had developed with the American Colonies. It was clear to members of the City

Council that Glasgow's wealth and position depended upon access to wider markets, especially on the Continent. The City, therefore, took a profound interest in communications, particularly where these affected the market for colonial produce or the supply of raw materials for the City's developing industries. This concern is evident in Glasgow's involvement in the deepening of the River Clyde and in the opening of the Monkland Canal, as well as the leading role played in the early moves that led to the opening of the Forth and Clyde Canal.[5]

Edinburgh, too, was anxious to ensure that the Capital's interests were safeguarded as the transport network expanded after the passing of the General Turnpike Act of 1751. The Trusts created by these and the many Private Acts were empowered to purchase land, make and maintain roads and control the traffic using the roads. Tolls were charged to finance repairs, as wheeled vehicles were liable to damage the surface. Although landowners were the most common proposers of turnpike trusts, since they saw the advantages of using carts to permit the movement of goods in bulk – lime, coal, farm produce and manufactures being the most frequent – others became interested. Investment came from merchants in the City of Edinburgh and, like its counterpart in the West, the City Council ensured that it was represented on any committee or trust in the immediate environs.[6]

Turnpike trusts, however, were of limited use for the carriage of heavy goods, as the roads were liable to cut up under the weight of the vehicle. Transport of these goods was slow and the commodities attracted high toll rates because the risk of damage to the road increased with the weight, thereby costing the trustees more by way of repairs. This led to complaints from merchants that transport charges increased costs by 'one day's work for a man and horse'.[7] Consequently, there was an incentive to look for a cheaper method of conveying heavy or bulky goods.

For over a century, there had been periodic discussion on the subject of building a canal through Central Scotland to connect the Atlantic to the North Sea. France had shown that a canal could be built 162 miles long and reaching 830 feet above sea-level when the famous Canal of Languedoc was opened in 1681, joining the Mediterranean Sea to the Bay of Biscay. It gave France a safe and easy sea-route from East to West and, as Colbert, Controller-General of Finance to Louis XIV, expected, it encouraged inter-regional, as well as international trade.[8] It was at this time that a canal in Scotland to link the 'East and West Seas' was first suggested, but came to nothing due to the enormous costs that were beyond the means of King Charles 11.[9] The major incentive that led to a reconsideration of the scheme was the work of James Brindley in England. In Glasgow, merchants noted that canals had been successfully opened for the Duke of Bridgewater in 1761, and extended in the following years from Worsley, to the great benefit of his coal interests. The success of this venture, and the commencement of work in 1766 on the Trent and Mersey Canal, indicated that a viable new transport system was now available and had the support of the merchant class. Scottish merchants did not wish to be left behind in such developments, and so contemporary surveys were carried out to discover suitable routes between the rivers Forth and Clyde. By 1767, a convenient route had been chosen that would open a new link with the Continent and, at the same time, provide a line of communication through the most densely populated and richest area of Scotland.

The Forth and Clyde Canal was the product of 'enlightened' thinking in two respects. It represented the response of the commercial classes to the developing economic life of the country: typified by Glasgow pressing for a small and cheap canal to further the export of tobacco to the Continent while providing a route for agriculture produce from the east to come more readily to the markets in the west. But at the same time,

Edinburgh, while sharing the economic interest in the canal, by urging the building of a large canal and attracting the support of the aristocracy and landowners of the east coast, gave an opportunity for the exercise of 'civic humanism' in the economic sphere to men of rank deprived of a rôle in Parliament: a process that was already well under way before 1760, but which the Canal was seen to typify by its aim of benefiting the community.

Early commentators were impressed by the magnitude of the undertaking and tended to stress the novelty, scale and complexity of the enterprise. Priestley, in his study of the canals of Britain, in 1831, makes use of the several Acts of Parliament that authorised each stage of the construction, between 1768 and 1820, to outline the development.[10] He explains the motives of the promoters as being the commercial interests of the merchants of Glasgow, making no reference to the part played by Edinburgh and the eastern burghs in the planning of the Canal, to the decision over the route, or to the controversy that divided the two cities. His main object is to trace the stages of building, and construction details and problems are only referred to in relation to the need to seek approval from Westminster. Cleland, writing in 1816,[11] gives more attention to both motivation and construction. He, too, is at pains to emphasise the concern of Glasgow tobacco merchants at the cost of land carriage, and their hope that the Canal would afford a means of reducing the cost of re-exported goods, while simultaneously opening up local markets along the route. His coverage of the actual building is confined to a chronological account of the work, stage by stage, but lingering on the size and scale of the Kelvin Aqueduct and the building of Port Dundas. He gives some detail about the organisation of the company, which indicates there was tension between Glasgow and Edinburgh, but does not elaborate upon it. Most writers have tended to follow this pattern and have concentrated on the constructional problems the builders encountered, like Rolt. Others, like C. Hadfield, have chosen to emphasise the financial problems that delayed completion of work for almost a quarter of a century.

More recently, there has been a tendency for historians to demonstrate that the Canal was an example of 'improvement' in transport, and as such, fitted the general spirit of the age. Professor Campbell, dealing with the navigation in his study of Scotland after the parliamentary union, suggests that transport changes were less urgent in the early years of the century since markets were sheltered from English competition. He demonstrates that as the century progressed, however, and freer flow of trade developed, the need for more efficient transport became obvious and Glasgow, which had benefited most from the Union giving trading rights to Scots, saw the great advantage to trade of linking the two major rivers and of avoiding the dangerous and costly journey round the North of Scotland. He briefly examines the financial and construction difficulties that had to be overcome, before considering in greater detail the economic impact of the Canal on Central Scotland. Similarly, Henry Hamilton is concerned to show that the building of the Canal fits into the philosophy of an 'improving' society. In his *Economic History of Scotland*, he sees the Dundas influence at work in the proposals for the Canal, suggesting that the project could be viewed as another means of expanding trading opportunities on the Forth, where ports were suffering a decline in the middle years of the century, and where the family estates were situated. In a later work, while still stressing the Canal as part of the 'improving' movement, Hamilton's emphasis changes, with the commercial aims of Glasgow being paramount, aided by support from like-minded businessmen from Bo'ness, with the role of Sir Lawrence Dundas being relegated to that of spokesman and advocate for these interests rather than any other. In both volumes, economic motives are proposed as the chief agents in advancing the cause of the Canal, and after a brief description of the building of the navigation, Hamilton concentrates on demonstrating the impact of easier, cheaper transport of heavy goods on the country.

The chief work on the Canal is by Jean Lindsay. In a detailed examination of the Canal, from origins to completion, she suggests that the commercial interests of Glasgow were the initial motivation that prompted an approach being made to Parliament, but that pressure from the eastern burghs, particularly Edinburgh, was so strong that the plans for a 'small canal' were withdrawn and replaced by a Bill for a much grander design. While accepting that the Capital was attracted by the increased opportunities afforded to trade by the Canal, it is suggested that the idea of a large public undertaking of general use and that was worthy of an advanced and sophisticated country appealed to many of the aristocracy living there. The slow development of the work is attributed to conflicting interests in Scotland, represented among the shareholders, which further complicated the financial and physical problems that bedevilled the Canal. The theme of rival schemes for the Canal is more fully developed in an article in *Transport History*, where she demonstrates that the early conflict between the two cities lingered on and was to influence the building programme right up to the opening of the Canal in 1790.[12]

The view that the Forth and Clyde Canal was the earliest, large-scale undertaking of the 'Age of Improvement' in Scotland is shared by many. Professor Smout sees it in largely economic terms, as a response to the changing economic base of the country and the need to provide a vital link to enable industry and agriculture to function efficiently and to develop further in the future. In his *Economic History of Modern Scotland*, Bruce Lenman agrees with this interpretation unreservedly. However, in a later work he argues that building the Canal was one aspect of a widespread 'improving' trend that influenced a whole range of social and economic activities in Scotland at that time.

A more restricted view is evident in the work of some other

writers. In *The Forth and Clyde Canal: A Kirkintilloch View*, the author, D.Martin, is chiefly concerned with the effect of the Canal on one local community, but gives a brief history of the development and building operations, particularly around Kirkintilloch. He is convinced that the motives that led to the work are explicable in terms of the economic interest of Glasgow merchants.[13] J.K.Allan, who also examines the Canal from a local point of view, that of Falkirk, gives more consideration to both the promotion and the difficulties. He explains the need for a canal in economic terms, but refers to other developments taking place in Scotland at the time and indicates that this goes a long way toward explaining why the idea bore fruit in 1768 and not earlier.[14] Examining the role of mercantile interests in the development of the Scottish economy, Professor Devine puts forward the view that the need for an easier route to continental markets was the prime consideration of the tobacco lords of Glasgow in advancing the cause of a canal between the Clyde and the Forth, and that this was part of a wider concern with communications that manifested itself in improving and deepening the Clyde and promoting the construction of the Monkland Canal.[15]

Whatever their differences of interpretation, there is a general consensus among writers that the Forth and Clyde Canal represents the greatest single venture undertaken in Scotland up to that time. All are agreed that it was essential to economic growth and that the merchant classes throughout the country were aware of this, but it is not universally agreed that the motives of the promoters were purely economic. There is more than a suggestion that matters of prestige and national pride influenced the final decisions on the Canal, and that the improving ideas of the age, expounded by teachers and philosophers, were put into effect by a generation for whom undertaking duties of civic importance had moved from the political to the economic sphere.

The present study examines the making of the Forth and Clyde Canal from the beginning up to the opening in 1790. The various influences behind the promotion of the Canal are assessed to discover which were most significant. The remaining part of the work assesses the reasons for the delay in completion, looking at construction and administrative difficulties as well as financial problems, before reviewing the Canal's early success, decline and subsequent revival.

2

Promotion of the Canal

The narrow waist of land that separates the River Forth on the east coast of Scotland, and the River Clyde on the west coast, is a natural line along which the country can be divided. This fact was first recorded by Tacitus as he chronicled the advance of the Roman legions of Julius Agricola into North Britain in the first century AD, and the line was marked by a series of forts, constructed to deter the marauding northern tribes from attacking the more settled southern territories of the Empire. Capitolinus, in his *Vita Antinini Pii*, noted that by 143AD Lollius Urbicus and his engineers had converted this line of forts into the northern frontier of the Roman Empire by building a turf and wooden boundary, the Antonine Wall. Urbicus was the first of a series of engineers to realise the advantages of building lines of communication along this, the narrowest landmass between the North Sea and the Atlantic Ocean. Roads, railway lines and the Forth and Clyde Canal were all to follow closely the route of his wall and its military road.

Military considerations had been the cause of building the Antonine Wall, and lay behind the first proposals for a canal in the seventeenth century. Impressed by the Languedoc Canal opened by his cousin, Louis XIV of France, and anxious to secure a safe passage for his warships to pass from the Atlantic to the North Sea, Charles II ordered a survey of the land between the two rivers. Twice in his reign, war had broken out with the United Provinces (modern Netherlands), and the King was concerned to find an easy means of moving ships between the 'West and East Seas', a concern that became more acute after the appearance of the Dutch fleet in the Thames in 1667.

The route between the Forth and Clyde was the shortest that could be found and was, therefore, likely to be the cheapest. The estimated cost of £500,000 was more than the impecunious Charles, whose income seldom reached the anticipated £1 million allocated by Parliament and never met his expenses, could afford and the scheme came to nothing.

After the Jacobite Rising of 1715, when the Government was forced into showing increased interest in Scottish affairs, notably in trying to disarm and pacify the Highlands, the scheme for a canal was revived and a survey was commissioned. In 1723, Alexander Gordon reported to the Barons of Exchequer that a canal was viable, but the *Scots Magazine* noted that Dullatur Bog and the River Kelvin presented serious obstacles to construction. Financial and technical difficulties seem to have deterred the Government and the idea sank into oblivion until 1741. In that year, William Wishart asked that a new and exact survey be carried out, and Gordon's assistant on the 1723 survey, William Adam, claimed that the magistrates of Edinburgh and Glasgow were considering raising £200,000 to finance a canal to link the two cities. Again the scheme came to nothing, due probably to the difficulties that such a sum, equivalent to the entire circulating medium of the Capital, presented to the intending participants in the two cities.[16]

During the 1740s, however, another, influential voice joined those who advocated the Canal. Daniel Defoe had journeyed throughout Scotland and in his *Tour Through the Whole Island of Great Britain* had been impressed by the likely economic advantages to be derived from 'a navigation from the Forth to Clyde: Irish trade could pass through Glasgow to London via the east coast'. But he was equally aware of the problem of financing such a venture and expressed the belief that funds would not be forthcoming until the demands of trade inclined people to provide the money. In this he foreshadowed, by some twenty years, the economic motives of many of those who advocated the

Canal, and anticipated their argument that easier transport of goods from east to west would not only be of advantage to the people of Scotland, but would provide a service for other parts of the United Kingdom as well.

By 1760, several changes had occurred in Scotland that favoured the cause of the Canal. The Jacobite Rising of 1745–6 had focused public attention on Scotland as never before, and there was a responsiveness in government circles to schemes to improve the economy. These could be funded from the income of the forfeited estates of Jacobites at little or no expense to the British Government. The intellectual movement known as the 'Enlightenment' was already well under way and provided a rationale for attempts to improve society; and the need to make the Highlands loyal to the House of Hanover provided the political will to experiment. The Commissioners for the Forfeited Annexed Estates were given the financial means of securing this loyalty, but were also charged with the task of 'civilising and improving the Highlands of Scotland'.[17] Thus, hand-in-hand with the need to end the traditional clan loyalty to the Stuarts went the aim of improving the economic prospects of the Highlander. Perhaps it was felt that if he had more to lose, he would hesitate to support the Jacobite cause in the future. However, it was easy for the loyal Whigs who ran the country to move from schemes that were particular to the Highlands to other grander designs that would serve the same 'improving' ends in the country as a whole.

Agriculture received the attention of the Commissioners, and the changes on the land stimulated an interest in easing the movement of goods to and from the urban settlements, particularly Edinburgh and Glasgow. The rapid growth of Glasgow due to the wealth created by the tobacco trade made that city a major market for farm produce, and it became the prime mover in schemes to improve the flow of manufactured goods from the city to markets, especially to London and the Continent. It is unlikely that this could have arisen earlier since it was not until the middle years of the century that the commercial class in Glasgow had accumulated sufficient wealth to be prepared to invest in schemes other than those that immediately affected their business. Twenty years earlier the finance had not been available for building the Canal, but by 1767 cash was not the problem it had been.

Not surprisingly, the idea of a canal was given more serious consideration then, and this was further stimulated by the Prime Minister, William Pitt the Elder, who suggested that public funds might be made available. Pitt had taken office during the Seven Years' War and was vitally concerned with the destruction of the French colonial trade, for which naval supremacy was of critical importance. A quick route through central Scotland that avoided the dangerous waters of the Pentland Firth would allow British warships to attack French shipping in the Channel and in the Atlantic with greater ease. Pitt's resignation in 1761 ended this prospect of finance, but had the effect of renewing interest in the scheme, an interest that grew as war underlined the benefits of a safe, easily defended passage for merchant vessels from one coast to the other.

The improving laird, Lord Napier of Merchiston, at his own expense, employed Robert Mackell and James Murray, engineers involved with the deepening of the River Clyde at Glasgow, to make a survey from Abbotshaugh on the River Carron to the Clyde at Yoker Burn in 1762. Although no action resulted from their report, the *Scots Magazine* noted that there began a period of intense activity that was, in the longer term, to result in the building of the Canal. Two years later, as a result of Lord Kames taking up the idea with the Board of Trustees for the Encouragement of Fisheries, Manufactures and Improvements in Scotland, the leading British civil engineer, John Smeaton, was invited to make 'an exact survey' of a route between the two rivers. His report, in 1764, suggested two possible routes for the Canal. One ran from the Forth at Carron, across the Bog of Dullatur to the Kelvin and through it and Yoker Burn to the Clyde. The other ran from the Forth north of Stirling to the River Endrick and into Loch Lomond, and thence, via the Leven, to the Clyde at Dumbarton. The first required 27 miles of artificial cuts, and to make a canal seven feet deep was estimated to cost about £80,000. The latter, by using natural waterways, reduced the amount of cutting to 17 miles, but because of its greater elevation required more locks, and this was estimated to be much more expensive. This was the last survey to suggest alternative routes, and pinpointed the Dullatur-Kelvin route as the most suitable. Thereafter argument centred on the terminus of the canal and its depth and not on its route.

That the Canal should pass through the most densely populated area of Scotland and link the industrial west with the grain-producing regions of the east was quickly pointed out as being of benefit to the community at large by those who urged work to start as soon as possible. Cheaper food supplies to Glasgow and easier transport of coal to Edinburgh, together with a greater flow of colonial and Irish trade that would use the shorter passage, were seen as the most obvious benefits that could arise, together with the prospect of providing benefits to communities along the route. The success of similar ventures in England was emphasised, according to the *Scots Magazine*, at a public meeting in Glasgow in February 1767, to show the practicability of the scheme, and investment was encouraged by showing how the toll charges on the Trent and Mersey Canal had already produced profits. Two weeks later, a correspondent of the *Edinburgh Evening Courant* showed a keen awareness of the importance of canals in England and Holland. He referred to the Bridge-water Canal and how it had aided the transport of coal, and to the wealth created by the 40 miles of inland navigation in the United Provinces. There was sufficient interest in canals for a book to be published in London in 1766 entitled *Advantages of Inland Navigation*. The author, Robert Whitworth, had been an assistant to the great canal builder, James Brindley, and was later to take over from Smeaton as Chief Engineer on the Forth and Clyde Canal.

Glasgow commercial interests were anxious to have an easy route for the re-export of tobacco, and were not very impressed by Smeaton's scheme to provide 'the noblest work of the kind ever executed'. As early as 1764, Robert Mackell had been approached by these merchants and invited to provide an alternative route that would bring a shallow canal to the city, but his suggested line through the Lanarkshire coalfield was considered too expensive, and the 240-foot summit at Bishop Loch presented construction problems. He was approached again in 1766 and the following year, the second time together with James Watt, and asked to investigate an alteration to the Smeaton line from Dullatur to bring the Canal into the city at the Broomielaw.[18]

Glasgow's economic interest can be gauged from the frequency with which merchants, individually or in organised groups, advocated the scheme, and it is probable that they were responsible for the petition in January 1767 to have a four-feet-deep canal constructed from Carron to the Clyde at Glasgow.[19] On 15 January of that year, the Dean of Guild informed the Council in Glasgow that a Bill was before the House of Commons and that the Merchants' House had appointed a committee to promote 'a canall or cutt betwixt the firths of Forth & Clyde to come near to the City at Broomielaw'; the Council agreed to appoint a committee of magistrates to assist the scheme.[20] The *Scots Magazine* reported a number of meetings in 1767, all attended by 'Noblemen, Gentlemen, Merchants and Manufacturers', to promote the canal, but most frequent reference is made to the 'Merchants & Traders'. Those who urged the building of the Canal were at pains to demonstrate the enormous increase in trade between the east and west coasts of Scotland during the century, in particular the movement of manufactures to the east and grain to the west, as was pointed out in the *Glasgow Journal* of 16–23 April 1767. Clearly, the Canal was seen as a means of reducing the cost of transporting goods between the two main cities of Scotland. Smeaton's estimates, and Mackell and Watt's amended routes, had all been based on a depth of seven feet, but as the Clyde was only four feet deep at Glasgow, the city moved that a four-feet-deep canal be constructed.[21] The deeper cut would cause more expense to the city, and make the transfer of goods at the city less necessary.

The *Glasgow Journal* recorded the presentation of the Petition to the House of Commons on 23 January and noted that the petitioners were from the commercial classes in the city. However, it has been pointed out that the motives of participants need to be differentiated: promoters were mostly interested in the economic advantages of the venture, but were obliged to offer 'financial advantages' to attract capital for costly and novel innovations from investors, who sought a profit. This was particularly the case in this instance where the undertaking was the largest ever attempted and without precedent in Scotland.[22] It is, therefore, possible to agree with Professor Devine's claim that the Canal was 'related to the tobacco men's need to link quickly with European markets'.[23] That is, the tobacco merchants were the promoters who saw economic advantages in the scheme. At the same time, Professor Smout's thesis that ventures like the Canal were a source of investment for landowners and merchants looking for a return for their surplus wealth, holds good, if the landowners and aristocratic members are seen as being 'investors'. The argument that Glasgow businessmen promoted the Canal is strengthened by the size of the subscription taken out on behalf of the merchants of the city, in the names of John Glassford and John Ritchie, leading members of the tobacco community: they subscribed £24,000 out of the total sum of £45,000.[24]

The financial benefits of the Canal, or a fear that it might divert trade from the established locations in the east, was certainly the reason behind the support given to the 'small canal' by the Carron Company and the merchants of Bo'ness. The traders of Bo'ness, a port of significance on the Forth, foresaw the potential of easier transport of goods to the west as threatening their survival as it would no longer be necessary to unload at the port for transshipping to the west. They therefore petitioned the Glasgow promoters of the Canal to make a cut from Carronshore to Bo'ness, of the same depth, four feet, as the main Canal. It could be argued that this support for the small canal was at least partly motivated by a desire to avoid the deeper canal that would have allowed sloops of 40 or 50 tons to pass through to Glasgow directly and bypass the port that depended on the shipping trade, i.e. that the motives of Bo'ness could well have been purely economic and protective and not too dissimilar to those of Glasgow. Self-interest was evident in the support given by the Carron Company. Samuel Garbett, founder and General Manager of the Carron Iron Company, was anxious to ensure that the eastern entrance to the Canal be located on Carronshore, near to the works, and at a point where the Carron Shipping Company could take advantage of the anticipated increase in traffic. He was also concerned about the effect of the Canal on the water supply to the Company.[25] Given assurance that there would be no interference with the water supply and a promise that Carronshore would be the eastern terminus of a four-feet-deep canal that would allow large vessels to pass through, Carron threw its weight behind the Glasgow scheme. The Council in Glasgow noted that the Company had taken 15 shares in the navigation, and empowered the Provost to invest £1000 in 'the canal betwixt Glasgow and Carron', in the name of the community.[26]

On 11 March 1767, Lord Frederick Campbell, MP for the City of Glasgow, together with Lord Mountstuart, was given leave to introduce a Bill for the construction of a four-feet-deep canal, and the Bill was given its first reading five days later.[27] The *Scots Magazine* reported the news on 19 March, and it seemed that at last the scheme that had been discussed so often in the past was about to be realised. But within days a correspondent became the first to raise questions about the scheme, demanding to know why Glasgow was in such a rush to have the Bill passed through Parliament before there had been full discussion, and especially before Smeaton's latest report was made public. In so doing, he began a two-year battle in the press between Edinburgh and Glasgow for control of the Canal which lingered on long after construction had begun.

3

Edinburgh v. Glasgow

The *Scots Magazine*'s correspondent challenged the 'small canal' on economic grounds, arguing that a deeper cut would reduce the cost of transshipping goods on the Forth and Clyde, and would also eliminate port charges and save time by permitting larger vessels involved in the coastal trade to pass from one coast to the other. He expressed surprise that Glasgow had not waited for the publication of Smeaton's report for the Board of Trustees for Fisheries, Manufactures and Improvements, which was 'due shortly'. East-coast interest should, he claimed, be consulted and the merits of each scheme debated. To allow for this, he suggested that the Glasgow Bill be delayed. On 28 March the *Glasgow Journal* carried a reply in such detail that it was published over two weeks. After outlining the protracted history of the scheme, the writer argued that the Glasgow plan should go ahead, as any delay could result in the abandonment of the enterprise. He criticised the cost of the more elaborate scheme and questioned whether the higher tolls that would be needed to defray the expense of building would really lessen the cost of coast-to-coast transport and if they would be of benefit to agriculture in the surrounding districts. He stressed that with 90% of the traffic being from the city, Glasgow should have the major say in the scheme.

On 6 April, however, the *Edinburgh Evening Courant* carried a letter that was to give a new turn to the argument. Up to that point, the discussion had centred on the economic advantages of either scheme, but the new correspondent introduced notions of prestige and honour. Smeaton's canal, he assured readers, would be a work in which all Scots could take pride, and he proceeded to launch a vitriolic attack on the 'small canal' and its promoters, who

> have affronted the public with a plan, grovelling as their own ideas; – a canal as narrow as their view, sordid as their dispositions, and shallow as their understanding: – a canal that has been justly called a CUT between Carron and Glasgow, – a ditch, – a gutter, – a mere puddle, – proper indeed for pedlars to puddle in.

He derided the idea that the chief function of the Canal was economic and urged considerations of 'magnificence and national honour', as well as the pleasure that it would afford ladies and gentlemen who viewed it. The many benefits of the 'grand canal' could, he assured readers, only be discovered after it was built. Commerce, he claimed, was of no concern to Edinburgh which was the seat of polite society and which should take the lead in all schemes and urged 'every nobleman and gentleman' to thwart the schemes of 'presumptuous traders'. Despite its venom and its stress on the rivalry between the two cities, the letter does place emphasis upon another element that was of significance during the 'Enlightenment', namely, the question of 'national and universal interests'. The building of the Canal was seen as a major work of great importance for the future development of the country, and it gave opportunity for the gentry to exercise their role as civic leaders. With the merchant class in Glasgow having taken the first step, it could be seen as this new class unsurping the traditional role of the gentry, and the criticism of 'pedlars', with the stress on the importance of Edinburgh as a cultural centre and home of the nobility, can be interpreted as a rallying call to the aristocracy to play its part in promoting ventures that were of concern to the country as a whole.

In the midst of this controversy, Smeaton published his report on 11 April. He advocated the line through Dullatur and a depth of seven feet to accommodate sloops of up to 40 tons. He estimated the cost at £100,000.[28] This proved that a large canal was viable and led to mounting pressure from east-coast burghs for the rapid implementation of the scheme in order to overtake the Glasgow Bill already before Parliament. While Glasgow complained that any alternative proposals would only result in unnecessary delay, the burghs on the east coast organised petitions to the Commons and held a series of meetings to bring pressure on the Glasgow promoter.

One writer, with poetic inclinations, expressed the view in the *Scots Magazine* (vol xxix) that Midlothian was intent on destroying the Glasgow Bill without any reason:

> There is a place where lie amass'd.
> Good bills propos'd but never pass'd,
> . . .
> This paltry place [so wills Mid-Lothian]
> The Glasgow Bill must make abode in.

But reasoned arguments were presented to show that greater economic benefits would follow from the deeper canal. The most obvious improvement would be in communications with the Continent and the Baltic – a point stressed by many writers at different times, in both the *Scots Magazine* and the *Edinburgh Evening Courant*, over a number of months. London and the towns of the east coast of Britain offered another source of increased trade if ships could pass from east to west through the larger canal. Emphasis was also placed on the benefits to east-coast towns of direct trade routes to the west. Not only would the fishing industry gain from improved access to markets, but the ports of Leith and Bo'ness would be able to trade directly with Ireland and even America. This last sounds like an inducement to persuade Bo'ness to end its support for the Glasgow Bill and lend its weight to the new proposals. In spite of the intention that the Canal should be a work of 'national pride', the arguments presented in favour of the large canal were mostly economic, and the conclusion seems inevitable that everyone was moved by these ideas and that other concerns were of secondary significance. The concern for pacifying the Highlands was evident in the suggestion that by making both the Highlands and Western Isles accessible to lowland influences, the Canal would encourage improvements there and bring stability. In the years following the French Wars of 1740–48 and 1756–63, it was not surprising that the security afforded to internal trade by the Canal should be mentioned as a factor in favour of having a large canal to take ships.

The benefits to the country as a whole won the support of the majority of the burghs, and the Convention of Royal Burghs in Scotland expressed the opinion that the Bill for the 'small canal' should be postponed to allow consideration of Smeaton's report and, at the suggestion of Midlothian, a series of country meeting were held to rally opposition to the Glasgow Bill. During April, meetings were held throughout Scotland, and only Kincardine (on 16 April) and Renfrew (on 22 April) dissented from the Midlothian call, arguing that the 'small canal' bill was so far advanced that it would be wrong to delay it, and they raised doubts as to whether the larger design would ever materialise. From the burghs of Edinburgh, Haddington, Selkirk, Linlithgow, Stirling, Perth, Aberdeen, Elgin and Banff there issued a series of petitions requesting that Parliament delay the third reading of the Glasgow Bill and accept a new Bill for a 'grand canal between the frith of Forth and the frith of Clyde'.[29]

When subscriptions were requested for the building of a 'great canal' on 1 May 1767, the first day saw £30,000 being promised, and the press reported that the target of £100,000 had been reached by the middle of the month. At a meeting of the shareholders in St. Alban's Tavern in London, it was agreed that Smeaton's plan for a seven-feet-deep canal through Dullatur be accepted and that Parliament be petitioned for leave to introduce a Bill. At the same time, the meeting asked Smeaton to provide fresh estimates for a canal capable of taking ships of 60 tons, and he reported in October that his original route, with a seven- to ten-feet cut would be suitable.[30] Edinburgh promptly urged that the depth of the canal be made ten feet, and pressed the Convention of Royal Burghs to seek financial assistance from the Commissioners for the Forfeited Annexed Estates.

Glasgow was persuaded to drop its Bill in return for a promise that a cut would be made from the new canal to the city and a payment of £1,200 as compensation for the costs incurred by the earlier Bill. A similar arrangement with the merchants of Bo'ness, and £300 compensation to them, allowed the new proposals to go ahead without opposition.[31] The Council in Glasgow had sent Provost George Murdoch to several meetings to look after the city's interests during discussions of the first Bill, and sent him with other magistrates and a solicitor to London to protect the 'interest and trade of this city' during negotiations over the new Bill. Application for a hearing from bakers and weavers with factories on the Kelvin was considered and a committee was appointed to note the problems of water supply that the building of a canal would create for these people. In November, as a result of the investigation, it was agreed that the promoters of the new canal Bill be requested to give security against any loss incurred by mills on the Kelvin as a result of the construction of the Canal and its water supply.[32] Glasgow Council still clung to its main concern of securing the economic wellbeing of its citizens that had been the reason for supporting the first canal Bill.

A new estimate was calculated by Smeaton to make the Canal deeper and provide additional cuts to Glasgow, Bo'ness, Stenhouse Miln Dam and Carron. The figure of £147,337 was much more than the promoters had anticipated and, with the refusal of the Government to give funds from the Forfeited Estates, considerable support would be required from the public. Nevertheless, the MP for Edinburgh, Thomas Dundas, introduced the Bill in the Commons on 14 May 1767, with the claim that it was 'of great advantage to the kingdom in general by reducing the price of land carriage'.[33] On 8 March 1768, the Royal Assent was given to Act 8. George III, c.65, authorising the making of a navigable cut from the River Forth to the River Clyde. It was generally supposed in Glasgow that work would start from the west end of the Canal near to the city.

The Act of 1768 empowered the 'Company of Proprietors of the Forth & Clyde Navigation' to construct a canal between the two firths, with a collateral cut to the City of Glasgow, with a view to 'Improvement of the adjacent lands, the Relief of the Poor, and the Preservation of the Public Roads'. Obviously, yet again, the needs of the economic life of the country are stressed and no mention is made, in the Act, of other reasons that had been advanced in favour of this particular scheme. The proprietors are named as the Dukes of Bedford, Buccleuch, and Queensberry, the Earls of Morton, Abercorn, Rosebery, Bute, Panmure, Fife and Catherlough, twelve Baronets, three Knights, the Lord Provost of Edinburgh, the Provost of Glasgow, and many gentlemen and merchants.[34] The presence of so many members of the gentry and aristocracy would seem to prove earlier claims that they were seeking an outlet for the exercise of their civic responsibilities that could not be satisfied because there was no Scots Parliament. But it is equally likely that these proprietors were responding to the economic need to divert surplus capital into investment schemes, and that the leading role assigned to them was a result of their social status and of the prestige they could bring to the company that been formed to make the Canal. A later amendment authorised an additional cut from Carron to Bo'ness, and the final route was defined as from the Forth near Carron, along a line through Bainsford, Bonny Mill, Dullatur Bog, Inchbelly Bridge, Cadder Bridge and St. Germain's Loch to the Clyde near Dalmuir Burnfoot, with collateral cuts from Blairdardie through Partick, over the Kelvin to Glasgow, and from Carron to Bo'ness. To fund the building, the *Glasgow Journal* reported, the proprietors were empowered to issue 1,500 shares, each at £100, and authority was given to raise a further £50,000 in the same way should more money be required.

The list of subscribers shows 127 names (see Appendix A), but absent are John Glassford and John Ritchie who had so generously aided the Glasgow Bill, although they may have been among those whose shares were held in the name of another person. However, Glassford at least was not uninterested in the new canal and sold land to the proprietors in 1771 – '35 acres, 1 rood and 27 falls of ground at Netherwood and Dullatur Bog' – for which he was paid £599.18/-.[35]

At the eastern end of the Canal, the Dundas family owned much of the land that the Canal had to pass through, and stood to gain by the venture, although no evidence of payments seems to have survived. The family also owned land at Castlecary over which the Canal passed, but again there is no record of any sale. It has been noted by some writers that the Forth and Clyde faced little opposition from landowners and was not subjected to restrictions to avoid inconveniencing large estates, as was the case with the later Union Canal. Indeed, the Glasgow interest, far from being disappointed by the failure of its own scheme, seems to have been sufficiently pleased with the concessions won from the new proprietors for an 'elegant entertainment' to be arranged by the Deacon Convenor and Deacons of the City to thank Provost Murdoch and his colleagues for their assiduousness in safeguarding the city's interest in the Canal Bill.[36]

The Carron Company, however, was not disposed to accept the Act, and Garbett commissioned James Brindley, Thomas Yeoman and John Golborne to make yet another survey to determine if there was an alternative, cheaper route. Their report in September 1768 indicated that marginal savings might be made by slight alteration of the western end of the Canal, but that Smeaton's line was the best available.[37] It was suspected by some of the proprietors that Carron was attempting either to revive the 'ditch canal' or to delay the start of work on the main Canal, and Smeaton was instructed to conduct the main design, but allowed to reply to Brindley by pointing out that the draining of 'Reay's Ford' on their line would be enormously expensive.[38] Carron, however, took out 15 shares in the Canal Company, but was prepared later to use these to exert pressure on the Company for the completion of the Carron collateral branch.

On 10 June 1768 Sir Lawrence Dundas cut the first sod on the Forth, distributed five guineas among the workers, and initiated work on the Canal. The economic ambition of the merchants and the investment of the landowners was, it seemed, about to bring to fruition a scheme that had been talked about for a century and which had divided Edinburgh and Glasgow, but had now brought them together in a common interest. No one could have foreseen that another generation would pass before the Canal finally linked the East and West Seas.

4

Obstacles to Construction

It took 22 years before the dream of a canal linking the Forth and the Clyde became a reality. Most of the reasons for the delay could not have been anticipated in 1768, as no one had experience of an undertaking on this scale, and the Company had to devise new and *ad hoc* methods of treating each new problem as it arose. There were three broad areas in which difficulties arose and caused delay. Structural problems that had not been anticipated at the inception of the Canal were to slow work down and cause the schedule to fall behind, and the delays were complicated by labour difficulties and archaeological finds. These delays increased costs and delayed the receipt of tolls that would have helped finance the venture. The paramount difficulty was finance. Smeaton's original estimates were found to be inaccurate, in terms of both work and expense, and were made worse by the delay in completion and collection of tolls. A nationwide lack of confidence in the financial system after 1774 aggravated the situation. To overcome these problems, there was a need to have a well-established and efficient organisation, but the Company of Proprietors of the Forth and Clyde Navigation was plagued by disputes over responsibility as Glasgow and Edinburgh vied with each other and London for control. Arguments among officers and members, and regular complaints from the Carron Company, took up much of the time of meetings, and it was not until the 1780s that the Company had settled into a system that allowed it to operate efficiently.

A major cause of the delay was that Smeaton had underestimated the difficulty of making the cut. His route is only 31 miles long, but it crosses 20 different rock formations:

whinstone, millstone grit, wenlock, basalt, granite, limestone and red sandstone. Even Mackell's altered route, while avoiding problems arising from using rivers as part of the route and reducing the number of formation to only 18, had to cross a major fault line near Renton. Of these, the whinstone areas created the greatest difficulty and required the use of explosives to blast a way through the rock before the navvies could start work. This was particularly slow work, and the need for much cutting and banking at Inchbelly and Cadder Bridge slowed work greatly, and prompted Mackell and the company surveyor, John Laurie, to propose changing the route over the Kelvin to avoid similar problems further west. This was authorised by a further Act of 1771.[39] Craigmarloch presented even greater difficulties; there, a very tough boulder clay, with a lot of gravel and rocks on the side of a hill, forced the contractors to build extra-high banking, and the work was further delayed by the discovery of a very hard whinstone sill that had to be blasted through.

Quicksand and mossy bogs created a second major physical obstacle to construction. At Dullatur, the bog was so deep that the contractors, Clegg and Taylor of Falkirk, were obliged to employ a special force of 50 workmen whose sole job was rebuilding the banks which kept sinking into the mud. This group sank 55 feet of earth and stones into the bog before the banks were made secure.[40] The bog also occasioned another, more unusual, delay when a number of bodies were discovered. Most of the corpses were those of troopers of General Baillie's Covenanting army who had fled from Montrose's forces at the battlefield at Kilsyth in 1645 and perished in the bog.[41] One trooper was found still mounted on his horse, with grain in his pouch and a look of terror on his face. The bodies had to be removed by archaeologists for examination before being given Christian burial, all of which stopped work on the site. Even when completed, the cut at Dullatur was still a source of problems, and residents claimed that their homes were being invaded by toads as a result of the draining effect of the Canal. The Company sent a representative to discuss the problem and he appears to have placated the inhabitants of the hamlet as there was no further reference to the problem.

Water supplies, particularly to the summit level, created another technical problem. The Act of 1768 had specifically forbidden the use of water from either the Carron or the Kelvin, where industrial developments were dependent on adequate supplies from the rivers. This made it difficult to ensure an adequate supply of water for the 16 locks on the eastern section, and the 19 on the western section of the Canal, essential for raising and lowering vessels 150–156 feet. The Company was obliged to build its own reservoirs to meet the need, and seven were constructed to give 15,958 lockfuls of water.[42] The critical supply came from Townhead Reservoir, known locally as Banton Loch, which covered 700 acres and completely submerged the battlefield where Montrose had defeated the Covenanting Army in 1645. This reservoir, however, created additional problems, and an aqueduct was built to carry water to the Canal at Craigmarloch, but it was found that a satisfactory flow could not be maintained. Several experiments were tried before it was decided to dam the Garrel stream and dig a channel through Colzium policies to increase the supply of water and provide an adequate flow to the Canal.[43]

Even with supplies from Townhead, Bishop Loch, Woodend, Gartsherrie, Johnstone and Possil Lochs, there was a need for a greater supply in 1785, when work began on the western section. The new chief engineer, Robert Whitworth, suggested that three reservoirs east of Monklands would be able to meet the demand, and would be less expensive, and far less dangerous than the alternative of converting Dullatur Bog into a reservoir.[44] His idea was discussed by a committee and it was agreed to meet the additional expenses: £4,466 for constructing the reservoirs, £562 for building an aqueduct from the east end of Monkland Canal to Cadder, and £2,407 to make a four-and-a-half-feet cut from Monkland to Hamiltonhill.[45]

The Carron Company, angered over the reluctance of the Canal Company to make the promised cut to Carronshore, expressed concern that the Canal was 'stealing' water from Carron. Carron's works manager, Charles Gascoigne, repeatedly warned the proprietors that their activities were affecting the water supply to the ironworks, but found that his complaints were ignored. Matters came to a head in 1775 when rubbish from the works blocked a stream and diverted water from the Canal and back to the works. Mackell accused the ironworks of deliberately interfering with the Canal's water supply and demanded that the obstruction be cleared. Gascoigne, with some justification, replied that the matter was a Carron affair and, according to Mackell's version, threatened to 'blow out the brains' of anyone attempting to remove it. Mackell took matters into his own hands and personally supervised the clearing of the rubbish to restore the water supply to the Canal. Gascoigne accused Mackell of being a 'public nuisance' and demanded that the Canal Company take action over his 'misapplication of power'. When the proprietors chose to do nothing to reprimand Mackell, relations between them and Carron, already strained on account of Gascoigne refusing to pay subscription calls, became worse, raising loud protests over the proposal to build a reservoir at Dullatur in 1785.[46]

Canal-building involved a considerable number of ancillary works. To meet the needs of construction, quarries were opened to provide building materials, reservoirs dug to hold the vital water supplies, aqueducts and tunnels constructed to carry the water to the Canal, and roads, rails and cranes had to be made to bring the materials to site. In addition, there were locks, gates, weirs and embankments, while paths, bridges and drawbridges were erected to allow passage across the Canal. On the Forth and

Clyde, this amounted to 39 locks, 35 drawbridges, 10 large aqueducts (including the massive one over the Kelvin), 70 miles of paths and a host of minor aqueducts and water channels.[47] Time, and money, was saved at a few places such as Twechar, where the course of a stream and an existing pathway were used for the route of the Canal.

All this activity required a large workforce, and the Company recruited locally at first, but found that many of the workers returned home at harvest time and left the work short of manpower, which delayed progress and resulted in deadlines not being met.[48] After 1770, Highland labour was recruited, but there was a similar pattern of absenteeism at harvest, and by 1778 there was competition from the Monkland Canal, which also required labour. Mackell accused Watt of 'stealing labour' and the Monkland's Company threatened to fine anyone from the rival canal trying to attract workers, although it was pointed out that wage rates were identical.[49] By 1785, Whitworth was employing Irish navvies (so called because they were building 'inland navigations') and some Englishmen, on the grounds that they were less likely to return home for the harvest. But he was mistaken, and delays occurred as the Irish labourers took casual work on farms at harvest as a break from the monotony of canal-building. The constant interruption of work due to labour shortages was a factor, although not a major one, adding to the other causes of delay.

The work of building itself was slow. Cutting the first sods so as to ensure the correct angle for the slope of the wall had to be carefully supervised and could not be rushed. Removal of the soil was laborious, becoming time-consuming when the bottom was reached, at which stage great care had to be exercised to ensure an even depth, and all the stones had to be removed. Once the cut was completed, it had to be made waterproof, and this was done by 'puddling', lining the bottom and side with a paste of clay mixed with gravel and sand, to a depth of between 18 inches and 3 feet. The mixture was made into a semi-fluid state by adding water and chopping it with spades to make a compact and homogeneous mass. This was spread in a layer over the floor to a depth of about 10 inches and trodden into place. After drying out for two or three days, a second layer was applied, and the process repeated until the correct depth of puddle was laid. Only then could the sides be given their coats of puddle.[50] On this basis, it would be 18 to 24 days after the cut was made before the section had been given its final lining, and in the case of Dullatur Bog, this would have been longer as there a greater depth of puddle would have been essential. Although this was time-consuming, it was cheaper than using cement which had to be imported from Italy, and was expensive, even though carried by Scots salmon boats that would otherwise be returning empty and which, therefore, gave favourable rates.[51] As a result, it was used only on important structures, like locks, where greater strength was essential. One spin-off from the use of puddle was that the wash from the vessels using the canal eroded the surface lining and produced a fine silt which was agitated in the otherwise still water and so prevented the establishment of weeds and plants that would have hindered the passage of vessels.[52]

Archaeological finds produced further delays, and although these were few, they added to the overall time problem. When the cut was being made at Castlecary, the contractors had to destroy part of the Antonine Wall. It was reported in the *Scots Magazine* that the Company ordered work to stop for several days to allow time for a thorough inspection of this important Roman site. A few months later, more finds were made at Auchenstarry and, again, work was held up to allow excavation of the site. The Canal Company took particular pride in presenting four Roman altars, unearthed at Auchendavy, to the University of Glasgow.[53] John Anderson, Professor of Natural Philosophy there, made use of these finds to detail the route of the Antonine Wall, and its structure. His papers, presented to the Glasgow Literary Society in 1770 and 1773, did much to encourage classical studies.[54]

The historical concern of the Canal Company is to be commended, although the damage done to the Wall is regrettable. The loss of several weeks' work could not be of any advantage to the Company, which could ill-afford to fall even further behind schedule. Yet this early example of rescue archaeology might go a long way toward explaining the motives of those proprietors who felt that they were engaged in an enterprise of 'national' importance, and the desire to preserve the past would accord with their ideas of civic pride and responsibility. Unfortunately, this concern for heritage did not last long, and on the western section of the canal, stones from the Roman fort at Cadder were used to line the banks of the canal, and near Old Kilpatrick it drove right through the bath house of a fort.[55]

5

Administrative Problems

Difficulties arose from the complex administration of the Company. With the Council meeting in London where it had access to Parliament, and committees in both Edinburgh and Glasgow to supervise the work on site, there were problems of communication and overall control. Minutes of all meetings were copied and sent to each centre together with copies of receipts, bills and accounts. This created a great volume of work for each meeting. For example, in 1773 an Edinburgh meeting had to deal with the minutes of eight other meetings, plus a letter from the London agent, and was also expected to audit the books.[56] With the primitive state of the roads, communication was difficult; the coach took 10 days to travel from London to Edinburgh, and there was no road between Glasgow and Edinburgh.[57] To overcome these problems, new procedures were developed. In London, the frequent contact with Westminster resulted in the Council adopting parliamentary procedures as a matter of convenience, and this gave rise to Members of Parliament being familiar with the working of the Company, enabling them to smooth the passage of Bills through the House.

When management of the construction moved to Glasgow in 1775, that city introduced a number of changes that not only eased the supply of information, but also ensured that control passed from Edinburgh to Glasgow, thereby guaranteeing that the commercial interests of the city would not be forgotten. This process of 'freezing out' the Capital had begun earlier. In 1774, London and Glasgow meetings had agreed to give Glasgow Customs House control of the whole Canal, and Edinburgh knew nothing of the matter until after the second reading of the

Bill. Glasgow, perhaps remembering how the Edinburgh interests had thwarted the original scheme for a canal, seems to have decided to gain its revenge by taking over the great canal and running it to the benefit of the merchants of the west. In 1782, after London had criticised the Edinburgh Committee for investigating problems in Falkirk, only two members turned up at the special meeting called in September to defend the action. Edinburgh had clearly lost its role as leader, and Glasgow had taken over. It was obviously of no help to the engineers and the labour force that the shareholders were engaged in their own private battle for power, and it must have hindered them in their quest to have decisions taken to expedite the work, waiting for agreement from all three centres involved and concerned lest disagreements between the centres caused the construction to be further delayed.

Patrick Colquhoun, former Provost of Glasgow, and leading administrator in the Company, encouraged the Glasgow interest to keep London supplied with all information that was needed, particularly an analysis of monthly revenues and end-of-year accounts. When the Government became a major shareholder in 1784, Colquhoun responded to the demand for regular and detailed information by developing new methods, including double-entry bookkeeping, accurate records of transactions and statistical presentation of information, together with pictorial summaries.[58] Gradually, however, ultimate control passed to London and Glasgow became an office of the Company rather than its centre.

Tighter control over administration was certainly necessary, as is evidenced by the continual bickering among the Company officers. Smeaton was an early critic of the way canal business was conducted, and complained of poor organisation due to the reluctance of the proprietors to employ sufficient officers to carry out the many tasks involved. In 1771, he even offered his resignation to save money, suggesting that the £500 be spent on hiring additional officers. The Committee refused his offer, but took no action to remedy the problems he had indicated.[59] Within a year, he was criticising the 'massive scale' of the construction, which he saw as increasing costs beyond the estimates. Again, the Committee took note of his comments, but did nothing about them.[60] It is possible that Smeaton was being unfair to the proprietors, as the structural difficulties certainly exceeded his expectations and required a more substantial construction than had been accounted for in 1767/8. It is also worth considering the comment of C. Small in the *Glasgow Herald* (11 December 1982) that most architectural work in the latter part of the eighteenth century was on a massive scale, corresponding to the style of the period. Nonetheless, the Company gives the impression of wishing to avoid conflict with Smeaton, but rather than explain the reasons for the estimates being exceeded, chose to note his complaints and, in a laissez-faire fashion, did nothing, which only served to worsen the relations between the Chief Engineer and the Company.

Robert Mackell, the Resident Engineer, emerges as the single most controversial figure in the history of the Canal. He seems to have fought a running battle with the proprietors, some of whom were keen to have him dismissed. In 1772, having dismissed two officers for being drunk on duty, he found that some of the shareholders were seeking to get rid of him, using the dismissals as evidence of his being unfit to hold an important position. With Smeaton's support, he overcame this problem, but the following year accused his deputy, John Laurie, and a surveyor, of causing the Company to pay more than was required following inaccurate measurements. Perhaps he anticipated a hostile response from Edinburgh, and he chose to write directly to London, where Colonel Masterton took up his complaint with the Council. Thomas Dundas of Castlecary was asked to investigate, together with Smeaton. Laurie retaliated by blaming the high estimates on Mackell and accused him of causing delays in the building. London resolved the deadlock by dismissing Laurie for 'contradicting the orders of the Resident Engineer'. The Scottish Committee reluctantly accepted the decision of London in April, but criticised Mackell's actions, while Edinburgh offered to re-employ Laurie. When Mackell accused Committee members of being drunk, his future looked bleak, but a re-measure of the disputed area proved he was right. Dundas's motion that Mackell be thanked for his 'attention to the interest of the Company' was accepted without dissent, albeit reluctantly. The contractors, Clegg & Taylor, disputed the new measurement and entered into a long and complicated lawsuit that was not settled until 1786. The shareholders, anxious to avenge themselves on Mackell, blamed him for the legal dispute and for the costs that this would produce. Unable to dismiss him, they made a final gesture by withholding his salary for almost a year after his death in 1779.[61]

Mackell constantly bombarded the Company with complaints – Carron stealing water, Monkland Canal stealing labour, the need for more officers, and the constant delays and rising costs. Moreover, his own records were haphazard, and the Company complained it could not distinguish which entries were purchases or when payments were due. In 1776, he was accused of using the Company stamp for private letters, and during the investigation a mob attacked his property. The Committee found him innocent of the charges but ordered him to keep proper records in future. This Committee was dominated by the Glasgow interest and they outvoted the Edinburgh members who had been campaigning against Mackell. Mackell's career had shown that it was possible to play the three centres off against each other, and he was able to use London and Glasgow at different times to overcome the opposition of Edinburgh. The emergence of Colquhoun's new administrative system in the late '70s, and the transfer of power to London after 1784, brought a chain of command that left major control of the Company in the hands of the Governor and Council in London, where there was less likelihood of the members becoming involved in the personal and parochial rivalries of the participants in Scotland.

6

Financial Difficulties

Finance was to be the main cause of the long delay in completing construction. It had been anticipated when the Bill was proposed that Parliament would authorise the Commissioners for the Forfeited Estates to release money to help fund this undertaking, which was regarded as being of great national importance. Government, however, felt inclined to leave it to private enterprise to find the £150,000 that Smeaton had estimated the work would cost.[62] The promoters had been cautious enough, as reported in the *Glasgow Journal* of 3rd–10th March 1768, to request power to raise a further £50,000 if required by authorising a further subscription.

The sum to be raised was enormous for a country as impoverished as Scotland in the middle of the eighteenth century, where, it has been estimated, the entire circulating capital of Edinburgh amounted to only £200,000, equivalent to the money required to build the Canal.[63] Scotland's only experience of a scheme of comparable magnitude had been the ill-fated Company of Scotland, which had lost the same sum of money in its attempt to found a trading colony on the Isthmus of Darien.[64] Half-a-century later, it is not surprising that investors in the canal venture were rather circumspect but the funds were subscribed readily enough, although there was no guarantee the call would be honoured. Against this background the reluctance of merchants to become heavily involved is more readily understood, as was, perhaps, their clinging to the idea of a smaller and cheaper cut. Significantly, the Carron Company was also concerned about cost, and this was the chief reason for Brindley being commissioned to make another survey in 1768, to discover if there was a cheaper route. This was used by the proprietors as a ruse to delay work on the Canal, and Brindley's suggested alteration to the Smeaton line was ignored until revived by Robert Mackell in 1770.[65]

Ward has pointed out that the motives of investors and promoters require to be distinguished, as they were not the same.[66] Promoters were particularly concerned to encourage and facilitate trade between the east and west coasts, while at the same time developing commerce along the route. Investors, on the other hand, were seeking a return on their money, although, in the case of the Forth and Clyde Canal, there was a genuine concern to be associated with a 'great national undertaking' that led to acceptance of both the 10% limitation on dividends and the inevitable delay before the Canal was operational and earning money.

When the 5% calls were made, however, not all those who had so enthusiastically subscribed were prepared to pay up. On the first call, no fewer that 203 subscribers did not meet the call and, despite reminders, these shares remained unpaid until 1773, when it was agreed that the remaining shareholders subscribe for a portion of the outstanding sum.[67] This reallocated the shares to 1297 shares of £115.13.2 ¼ d.[68] Forrester had used the Minute Books of the Canal Company and the Record of Payments by the Glasgow Agent of the Council to identify the 13 calls. This table shows clearly that even a regular payer like the City of Glasgow did not always respond by the due date, thereby causing the Company to run for some time short of the required funds.[69] These calls can be seen to represent the different phases of building the navigation. The earliest ones raised the cash to pay the expenses of having the Act of Parliament passed and paying £1200 compensation to Glasgow and £300 to Bo'ness for their small canal Bill.[70] In addition, this first call had to raise the funds to prepare for the making of the first cut in June 1768.

Some of the investors were reluctant to pay. David Loch, who in his *Essays on the Trade, Commerce, Manufactures and Fisheries in Scotland* (1778) had lavished praise on the 'public spirit' of those who had financed the Canal and urged others to subscribe to 'one of the noblest undertakings recorded in history', was found to be £475 in arrears in 1780. At about the same time, the Company was finding it difficult to enforce calls in England, where the law was different and made little provision for action against defaulters. It was agreed that the system to be followed would be that of a polite reminder that payment was overdue, and if this did not result in payment, the London banker, Drummond, would visit the defaulter bearing a receipt. In the case of Lord Scrafton, this failed to produce the funds, and the Company took legal action that resulted in the nobleman's executors being ordered to pay the £1000 that was due, and the expenses of the action in King's Bench Division as well.[71]

Recourse to the courts was agreed at a meeting in London on 17 February 1773, although it was hoped that the threat of legal action would be sufficient to produce the arrears.[72] The Company did not like the expense of lawsuits, and was afraid that if it lost the case, it would have to meet the costs, as happened in 1787, when the court decided in favour of Henry Isaac's executors. A second reason for the reluctance was the adverse publicity that would arise if the person sued was a prominent landowner. It was agreed that a more effective means of encouraging prompt payment of call might be the threat to charge interest on the outstanding sum. By 1781, this had not produced the desired affect, and arrears of £14,300 were recorded. Two proprietors in total arrears were struck off the list of members, but Lord Elphinstone was sent a polite reminder that his subscription was overdue. By March 1785, it was announced in the *Scots Magazine* that the Company had decided to charge compound interest on any arrears.

Perhaps the chief defaulter was the Carron Company, which used its 15 shares to bring pressure on the Canal Company to make the cut that had been promised to the works. Carron's earlier support for the Glasgow Bill had been in the interest of easier access to the west coast for the iron goods produced at Carron, and to give the Carron Shipping Company a virtual monopoly of carriage on the east coast. Only the promise of a cut to Carron had persuaded the Company to support the 1768 Act and subscribe for 15 shares, held in the names of John Adam, John Clerk of Eldin, and the Company Manager, Charles Gascoigne.[73] Gascoigne had sent a sketch of the cut that the Company required to the meeting of proprietors in London in 1768, and was concerned that the meeting placed a low priority on the cut.[74] He and Francis Garbett decided to withhold payment of the calls to coerce the Canal Company into making the Carron cut. In 1771, Gascoigne again asked that a cut be made from Lock 4 at Dalderse to Carron, but was informed by Smeaton that this would cost more than the estimated £5725, as the line proposed would require a new Act of Parliament. Instead, he suggested a cut from Lock 5 at Bainsford through land that the Navigation Company had been authorised to use by the Act of 1768. Carron won over two of the proprietors, Lords Rosebery and Elibank, who urged that Carron be allowed to make the cut from Bainsford at its own expense and recoup the money from lock dues. The Canal Company rejected any suggestion that its legal right to collect tolls be surrendered to another party. Carron sought legal advice over this 'breach of contract' and refused to meet calls on the subscriptions until the Company's grievances had been redressed.

There is a hint in the actions of Carron that the Company wanted to delay construction of the Canal to gain its own ends. If the venture were abandoned, Carron, with its Glasgow allies, could have revived the small canal, which would have given Carron Shipping a monopoly of the coastal trade and increased its business by carrying the goods being transshipped to the small barges. Should the larger canal proceed, there were advantages in having the eastern terminus at Carron where the Canal Company would need to assume responsibility for maintaining the passage to the Forth, thus saving the Carron Company expense.

In 1773, the London Council appointed the Duke of Queensberry to negotiate a compromise with Carron. He offered to have a cut made through the loops of the River Carron to link the Canal with the works, if at the same time the Carron Company brought its subscription payments up to date. This compromise agreement was, however, ignored in Scotland, where the first priority was to push on to the west coast and Glasgow, without diverting funds, or labour, to a cut for the sole benefit of Carron, and Smeaton was instructed to continue with the work on hand.[75] As a result, Gascoigne thereafter waged a campaign against the Canal Company, accusing it of stealing water from the works to supply the Canal, while he continued his legal battle over the breach of contract. To force concessions from the Canal Company, no further payments were made and the affair dragged on while the financial crisis deepened and the funds became exhausted. When construction was brought to a halt, there was no prospect of Carron getting the cut it wanted, and it was not until 1783/4 that work recommenced. However, by this time the Canal Company had decided that no cut would be made to Carron, and it ended the protracted dispute by exercising the powers conferred by the Act of 1771 (11 George III, c. 62) and it informed the Carron Company that the shares had been forfeited for failure to meet the calls.[76] Carron, urging its own interests too forcibly, had angered the Canal Company and, by pressing its claim as a priority, had alienated even its Glasgow allies, who feared that the eastern extension might jeopardise the cut to Glasgow and affect the opening of the Canal on the west coast itself. In a clash of interests, Glasgow had the greater influence, especially after 1775 when management of the enterprise was based in the city, and its interests were inevitably to prevail.

Aerial view of the Forth and Clyde Canal at Banknock. This demonstrates the importance of the area for transport from early times. The Canal (left centre) turns right after Wyndford Lock to bypass Banknock and go through Castlecary (centre right) before going under the road, A80/M80 (centre, right to left). The Glasgow to Edinburgh railway line (centre right) crosses the viaduct at Castlecary, and the outline of the old Roman military way is defined by the line of trees on the extreme right. (*M Dowds, with thanks to Scotia Helicopters, Cumbernauld*)

Symington's Plan for a Paddle Steamer, 1789. William Symington (1764-1831) had already patented a steam-engine in 1787, and drew this sketch of a canal boat that could be powered by steam. Financed by Patrick Miller, he introduced such a steamer on the Forth and Clyde Canal, achieving a speed of seven miles an hour. (*History Research Centre, Callendar House, Falkirk*)

The *Charlotte Dundas*, 1802. Symington developed a new paddle steamer, which he named after the wife of his backer, Lord Thomas Dundas. With a single paddle mounted in the stern, it was different from other vessels and, although it pulled two heavily laden barges over 19½ miles against a strong headwind, the scheme was abandoned because some of the shareholders feared that the wash would damage the Canal's bank. (*William Patrick Library, Kirkintilloch*)

TOP.

The Kelvin Aqueduct. Designed by Robert Whitworth, the aqueduct carried the Canal over the River Kelvin so as not to interfere with the water supply to industries on the river. At 70 feet high, it was one of the most imposing structures on the Canal, and the sight of sailing boats passing over the river impressed Glaswegians. (*MS Murray 636, University of Glasgow*)

BOTTOM.

The Kelvin Aqueduct, Maryhill. This view shows clearly the four arched spandrels on each side of the 400-foot-long aqueduct, as well as the later housing and industrial developments that sprang up alongside the Canal. (*William Patrick Library, Kirkintilloch*)

Launch of the SS *Nelson* 1893. Shipyards were developed at several points on the Canal to build the vessels required by the transport companies. The earliest were at Kirkintilloch, where J. & J. Hays was one of the first. The photograph shows the launch of the *Nelson* from the Hays' yard. The vessel is being launched side-on to the Canal, and the event has attracted a large crowd.
(*William Patrick Library, Kirkintilloch*)

Aerial view of Kirkintilloch Basin, 1930, illustrating the extent of the development that centred on the Canal. Road, rail, housing and industrial sites straddle its banks. Dating from the period of economic depression, there is a lack of vessels on the Canal and of work in general.
(*William Patrick Library, Kirkintilloch*)

TOP.
Port Dundas, c. 1890. Where the Forth and Clyde and Monklands Canals met at Port Dundas became one of the busiest harbours on the waterways. Large sailing vessels carried passengers as well as cargo, while the small scows and lighters handled bulk cargoes like coal, as shown here. (*William Patrick Library, Kirkintilloch*)

BOTTOM.
Fishing boats at Kirkintilloch. A major user of the Canal was the fishing fleet, which each year used the short, safe route between the white fishing and the herring fishing grounds. The photograph shows fishing boats on their way through in the 1950s. The number of ships declined over the years, and had fallen to 98 by 1956; the appeal of the fishing industry to retain the Canal was ignored by government and its closure was ordered. (*Provost William Fletcher, William Patrick Library, Kirkintilloch*)

Horse-assisted steam lighter, 1930s. Lighters were small vessels used to transport bulk cargo. Although steam power had been in use for a century, lightermen preferred the horse. Here a horse is being used to take a steam-powered lighter through the sharp bend known as 'The Deil's Elbow' near Auchinstarry. (*History Research Centre, Callendar House, Falkirk*)

BOTTOM.

Union Inn, Camelon. Travellers undertaking the 13-hour journey between Edinburgh and Glasgow made use of this famous inn to have a meal while their passenger boats passed through the 11 locks that lowered the vessels the 115 feet from the Union to the Forth and Clyde. The inn is still operating today and is close to where boats leave to take visitors to the new Falkirk Wheel, which replaces the long-demolished locks. (*Author*)

TOP.

Falkirk Wheel and Visitor Centre. Opened in 2002 to complete the Millennium Link, this unique lift was built to re-connect the Forth and Clyde and Edinburgh Union Canals. Opposite the glass-fronted visitor and exhibition centre, the Wheel raises up to eight vessels in the two 300-tonne gondolas, the 115 feet in only 15 minutes. (*Author*)

BOTTOM.

The Wheel in motion. Boats entering the two gondolas displace their own weight of water and these are perfectly balanced at all times. This balance, together with the precision engineering of the Wheel, means that they can be transported the 115 feet using only ten 7kw. motors – about as much power as is used by eight toasters. (*Author*)

The Royal Opening, 24 May 2002. 234 years after her royal predecessor gave approval for the making of the Forth and Clyde Canal, Her Majesty the Queen formally opened the re-connected waterways. During her visit, she inspected the Falkirk Wheel and is seen here in the Visitor Centre with the wheel towering over the windows behind her. (Michael Gillen, *Falkirk Herald*)

Canal building required a heavy investment of capital and considerable patience on the part of the investors as there could be no return from tolls until the navigation was opened, usually between five and ten years. The Forth and Clyde Canal took 22 years in the building, although revenue did come from the Grangemouth to Kirkintilloch section which was opened to traffic in 1773. Up to that time, the calls on subscription had to raise the money to pay for the construction.

The first two calls were used, almost exclusively, to pay for the cost of the legislation authorising the Canal. To acquire land to build the waterway across Scotland, the proprietors were obliged to seek a private Act of Parliament. After surveys were completed, the route and estimates of the cost were passed on to lawyers, acting for the promoters, who drew up a petition to the House of Commons. Once the House had consented to the petition, a Bill was drafted, Members of Parliament were lobbied by advocates for the Bill and public meetings organised to explain its purpose and win support. The expense of the first phase was considerable, and in the case of the Forth and Clyde Canal would have been more than the £1200 that it had cost Glasgow to present its small canal Bill, although detailed figures do not seem to be readily available.

With Parliamentary approval finally secured, the proprietors were faced with the task of acquiring the land needed for the construction. A detailed breakdown of the price paid for land is not available, but estimates can be made from the Dispositions held in Falkirk Museum.[77] Taking an average of the price of land in the Dullatur to Kirkintilloch area, a figure of around £17 per acre emerges, which is less than half the price being asked on the outskirts of Glasgow, where £45 an acre was paid. This enormous increase in land prices may go a long way toward explaining the rapidity with which the Company ran out of funds as it built westward, and the keenness of the proprietors to sell the unused land at Port Dundas in 1790. This last was a means of earning money by making the land available for house building, although it was seen by contemporaries as an attempt to produce a 'pleasing effect to the eye'.[78] In comparison with later ventures, the Forth and Clyde faced comparatively few problems over land purchase. One explanation for this was the system that was used in land purchase. Land was valued and owners trusted in a fair valuation, interest payment being made after work had started, with an appeal to arbiters in the event of disagreement. As a result, the proprietors were able to acquire land and commence operations before a price had been fixed. The Dispositions in Falkirk show 100 entries for land acquired in 1771 and paid for between 1772 and 1779.

The Navigation Accounts Books for 1803 show payments for land acquired in 1768, 1772, 1789 and 1796, and the balance due on land purchases was not finally cleared until 1816, 26 years after the Canal was opened from sea to sea.[79] Only one serious attempt to challenge a valuation can be identified. A Glasgow merchant named Robert Lang sought legal advice when the Canal Company refused to pay £50 an acre for land at Kilpatrick at the western end of the Canal. His lawyer argued that the land was a bleachfield and that their client was entitled to compensation for the loss of such a valuable property. Lord Genlee received a number of submissions from both sides and presided over a bitter argument that descended to an attack on the worth of some of the witnesses – Lang claiming that the weavers called by the Canal Company were not competent to assess the utility of land as bleachfields. In 1780, Genlee decided that Lang was due some compensation for the loss of the land and ordered the Company to pay him £657–9/– for the 15 acres, i.e. around £44 per acre.[80] This action was exceptional, and most landowners were content to allow an asessor to fix a value on the land.

Construction costs were high, despite efforts to keep wages to a minimum. Smeaton was given a salary of £500 a year to supervise the work, and this was later criticised on the grounds that he only visited the site periodically, leaving the day-to-day supervision to the Resident Engineer, Robert Mackell. In the early years, he seems to have had difficulty in getting on with the proprietors and offered his resignation in 1771, but this was rejected. In 1773, however, he resigned in a disagreement over how the work was to be executed, stressing that it was more ambitious than he believed necessary.[81]

The *Glasgow Journal* reported in March 1768 that Robert Mackell had been appointed Resident Engineer at a salary of £315, and was in charge of the operations on site. Mackell felt his position to be insecure after the resignation of Smeaton, who had defended him loyally in a series of clashes with the proprietors, and when work came to a standstill in 1775, he informed the Company of an attractive offer from Russia, but agreed to stay with the Canal Company if his salary was guaranteed for five years. Although suspicious that the Russian offer was not genuine, the proprietors yielded to his demand and he remained in charge until his death in 1779.

No successor was appointed until work resumed in 1784, when Robert Whitworth, a pupil of James Brindley, was made Chief Engineer with the same salary as Mackell, although in the following years the proprietors repeatedly and publicly praised his constant attendance on site.

The largest single item, in financial terms, was the wage-bill for the workforce. By 1770, there were 1048 men employed on the construction, and even after the initial phase of cutting, the labour force settled at around 700 men.[82] The Company had set the wages at 'not more than 10d. a day' and in 1769 were in receipt of letters from Edinburgh expressing concern that the navigation might lead to an increase in day-labour rates in the Capital.[83] The Company agreed to keep wages in check, but did authorise an extra 1d. a day for work in winter. This resulted in an outlay of between £30 and £40 a day in wages alone, although a subsidy of 6d. a head for six months for every young person employed went some way to easing the burden in the first two years. By 1773, it was estimated that labour costs to make a cut from 'Logie Water to the West Sea' would amount to some £4500.[84] In addition to the wages of the navvies, the Company had to hire quarrymen, supervisors, carpenters and masons, and Smeaton claimed these last two were paid more than in England.[85] These ongoing costs, together with the slow payment of subscriptions, made the financial problems critical and explain in large measure how the Company ran into cash-flow problems that resulted in work having to stop in 1775.

A system of sub-contracting for the digging was operated from 1770, with the sub-contractor being allocated a specified length of cut to make and being given $3\frac{1}{4}$ d. and $\frac{1}{2}$ d. per cubic yard of 'stuff' removed. This work was closely scrutinised by the engineer, who had to accurately assess the amount of soil removed as there was always the possibility of error, intentional or otherwise. The dispute between Mackell and Laurie arose over the former's claim that his deputy had been careless in making measurements and that this had cost the Company money. An independent survey showed that there had been an over-estimate of 69,859 cubic yards, although the sub-contractors, Clegg & Taylor of Falkirk, refused to repay the £1000 claimed by the Canal Company and entered into the legal action described earlier. In general, however, wages paid by the sub-contractors were fixed by the Company, which also kept close scrutiny on the progress being made, and problems were surprisingly few. When work recommenced after the delay from 1774 to 1783, sub-contracting continued on the same terms as before, and the last stretch was made to bring the Canal to Bowling. In spite of strict supervision, the time taken to complete each section was longer than had been anticipated, with the result that costs were rising. Evidence of the rate at which costs rose can be deduced from the estimates for making the cut from Carronshore to Bo'ness that had been agreed in the Act of 1768:

In 1768 cost estimated for the cut was £8,000
    1783   do,   £12,000
    1789   do,   £17,763.[86]

As a result of the increasing expense, the scheme was finally abandoned in 1789. The figures also indicate another major reason for the Company running into financial problems.

In 1768, Smeaton had proposed dividing the Canal into three sections and having all three under construction simultaneously. The Company had rejected the idea as too expensive, particularly as three surveyors, three foremen and a large workforce would be required.[87] Perhaps if they had been prepared to lay out the additional money at this early stage, the Canal would have been completed much sooner and the savings effected by not having to meet the rising costs in later years would have compensated for the heavy initial outlay.

Maintenance was another major item of expenditure that continuously ate into the available funds. By the time the Canal was opened as far as Kirkintilloch, the annual cost of repairing the completed section was £1961.4/-. This included the salary of toll-keepers and a small workforce which made good any damage done by vessels using the Canal. As the cut was extended, it was obvious to Patrick Colquhoun that the costs would rise considerably. He proposed that maintenance and operation of the Canal should be sub-contracted as a means of reducing costs to the Company. A private Act was secured in 1790 (30 George III, c.73) which extended to the Company rights similar to those of turnpike trusts and allowed the leasing of tolls.[88] Colquhoun's rise to become an influential figure of the Committee led to the adoption of a number of new strategies that made finances less burdensome and which gave full information to Council in London.

An intermittent source of expense, but one which could not be estimated in advance, was the cost of injuries to workers. In November 1769 Mackell reported to the Company that several men had been injured as a result of accidents during construction and some had been disabled. He was instructed to give them some 'small comfort during their recovery' and informed that the Company would pay for medical attention.[89] By 1770, however, the cost of medical treatment had risen dramatically, reflecting the difficulties and hazards of the work, and the following year the Company received bills of £64 from surgeons in Falkirk and £17 from others in Kirkintilloch for treating injured workers. When the western section was started in 1784, there was a need to use explosives to blast through the whinstone rocks and this led to an increase in the number of accidents. As a result, the Company decided to make a special arrangement to have their workers treated at the recently built Glasgow Royal Infirmary. By 1789, it was agreed to make a subscription of £200 to the hospital.[90] Paternalistic care for the workforce was, however, somewhat moderated by economic considerations, and when a man was injured at Falkirk in 1790, he had to await the next vessel going to Glasgow before being given medical treatment, which was free in Glasgow but which would have required a fee to a doctor in Falkirk. No record of what the Company spent on medical attention is available, but it was probably close to £1000, which represented a cost that had not been allowed for from the start.

Revenue after 1773 was seldom more than £5000 from tolls, mainly on the carriage of corn and timber, and did not increase until the Canal was fully operational. After completion, there was a steady increase from £9764 in 1787, to over £11,000 in 1791, soaring to £22,000 in 1797, this last largely due to the French Revolutionary War making the safe route through the Canal more attractive.[91] It was the lack of income from 1773 that precipitated the major financial crisis that led to the virtual abandonment of work on the navigation for a decade.

In the early stages of construction, the Royal Bank of Scotland had provided overdraft facilities and obtained the best rates of interest for a work that it regarded as being of 'national concern'. The Bank charged only 4% on loans, in return for the Canal Company using Royal Bank notes, but securities were requested from the proprietors. The Dundas family gave their personal bonds to allow the Bank to discount bills from the Company to give it immediate cash, which would not be repaid until the sixth call, due at the end of 1770. Thanks to the overdraft arrangements, the Company was able to continue work on the Canal, but, with no income from tolls and the subscriptions not always being honoured, a debt was soon accumulated. By 1772 the Dundas family was liable for £19,600 that was owing to the Bank, while the Governors in London calculated that the funds needed to complete the work 'from the Logie Water to the West Sea' would amount to £102,000. They broke the figure down to show how it arose: £19,600 was owed to the Bank, £4500 was to be paid on salaries, and £7800 for land purchase and, with subscriptions bringing in only £39,000, the Company required to raise about £70,000 to continue operations.[92] The Company tried to solve the financial problem by the issue of a bond against income from tolls that were expected when the Canal was completed, and offered 5% on these bonds. This created two problems. Firstly, 5% interest represented a heavy charge on the Company, especially as it was not possible to calculate when the bonds would be redeemed. In the event, work ceased on the Canal and the bonds resulted in even more severe financial problems for the Company. More significantly, the issue of bonds against future income, based as it was on one of the methods that had been used to prop up the Ayr Bank, was psychologically unsound. The collapse of Douglas, Heron & Company two years earlier in 1772 had devastating effects on

Scotland. The immediate collapse of 13 private banks in Edinburgh had ruined confidence in all financial institutions. For the
Canal Company, the situation was more acutely embarrassing as
it shared with the Ayr Bank a number of prominent promoters:
the Duke of Queensberry was chairman of both ventures and the
Duke of Buccleuch was a leading member of the Councils of
both.[93] With confidence gone, there was understandably a
reluctance to risk money in ventures like the Canal, and efforts
to rouse interest failed.

Simultaneously, Glasgow, the strongest supporter of the
Canal, was suffering an economic crisis. The dispute with the
American Colonies had begun to affect the tobacco trade, and
this had seriously undermined the financial position of the city
and removed one of the chief reasons for its enthusiasm to have
the Canal built. In the ensuing depression, merchants were
reluctant to venture capital on the Canal, which had been seen
primarily as a means of re-exporting tobacco from the west coast
to the Continent. This, however, is only a partial explanation for
Glasgow's failure to respond to the appeals of Sir Lawrence
Dundas and the Edinburgh agent of the Company, George
Chalmers, to raise an additional subscription in 1776. Glasgow
insisted that its original design of having the western terminus in
the city be agreed as a priority before it would consider making
funds available. Money does not seem to have been the major
concern, as the merchants were able to raise a private subscription in 1776 to pay for the extension of the cut from Stockingfield to Hamiltonhill to bring it nearer the city. It has been
suggested that the desire to have the Canal enter the Clyde at
Glasgow was an example of financial stringency at the time; the
same stringency that was behind James Watt's urging a cheaper
route for the Monkland Canal.[94] Certainly, the city proponents
of the Canal were more cautious than their Edinburgh counterparts in that they had advocated the 'small canal' in 1767, and
only agreed to the larger design with reluctance, nor were they

enthusiastic about schemes to deepen the Canal to accommodate larger vessels. This apparent caution could, however, be equally well attributed to the merchants' desire to secure the Broomielaw as the point of entry to the Clyde as a means of increasing the prosperity of Glasgow. Their rallying to fund the cut to Hamiltonhill shows that the reluctance was not due to lack of funds, but rather to a lack of will to finance the rival entry.

7

Government Aid

From the start, repeated requests had been made to the Government for financial aid; the last before work ended was in 1776. These appeals were rejected on the grounds that public funds were not to be applied to private projects like the Canal. When Sir Lawrence Dundas and George Chalmers met with refusal in 1776, and failed to secure an additional private subscription, new and more careful approaches to the Treasury were planned. The Dukes of Argyll, Buccleuch and Queensberry, together with the Lord Advocate, Thomas Dundas, met with Lord Shelburne at the Treasury in 1783 to solicit his help. The following year they had meetings with the Prime Minister, William Pitt the Younger, and the Treasury Lords to arrange financial assistance, and by March were assured that their request would be favourably considered.[95] Although the war with France had ended in 1783, it had shown that a canal would serve a useful purpose at such a time by providing a safe route between the east and west coasts of Britain. As the Government was considering reselling the annexed estates of the Jacobites, the Canal Company was able to make a sound tactical move by advancing £200 toward the cost of the Disannexation Act, in return for which it hoped to secure a share in the proceeds. For whatever reasons, the Government decided to abandon the mercantilist ideology that had stopped it making money available earlier, and instructed the Barons of the Exchequer in Edinburgh to give £50,000 to the Canal Company from the Forfeited Estates Fund.

The Act of 1784 (24 George 111, c. 55) gave aid in order to provide safe transit for vessels across Scotland, to ease access to the Western Isles and to avoid the dangerous journey via the

Pentland Firth. It was agreed that the dividends due to the Exchequer were to be used for road- and bridge-building, and the Canal Company was required to deliver an abstract of its books every year to the Treasury. This last condition added impetus to keeping accurate records, and produced an efficient copying system. Patrick Colquhoun initiated a system of double-entry bookkeeping and a General Account Current to enable efficient and understandable information to pass between Glasgow, London, the Scots Exchequer and the King's Remembrancer. As a result, the Company's accounts came to be recorded in the same manner as the Government's.

The Exchequer became a proprietor of the Canal Company and was shown in the Accounts as owning 2999 shares.[96] But as no dividends were paid before 1800, the Government investment did not yield any profit, and when repayment was made in 1798, it agreed to forego all interest payments and accepted the £50,000 capital, which was then used to finance the Crinan Canal and Leith Harbour. The Government, obviously influenced by the success of the Forth and Clyde venture, seems to have changed course and decided to help fund similar ventures.[97]

State aid enabled the Canal to be opened from sea to sea, but at the cost of a change in the power structure of the Company. After ousting Edinburgh, Glasgow had been the centre of operations from 1773, and had established very good working relations with the Council in London, often informing Edinburgh of decisions after they had been agreed between the two other cities. With the Government a major shareholder, however, it was natural that London should play a more prominent role in decision-making and that Glasgow should end by being sent instructions on how to conduct affairs so that public funds were protected. Yet the Canal and Glasgow left a lasting impression on Government. George Drummond, a Commissioner for Examining at the Treasury, and related to John and

Adam Drummond, Proprietors of the Canal Company, introduced the system of cash accounting at the Treasury because he
was impressed with the way it had been practised by the
Company.[98]

By 1786 the Canal Company had overcome all the major
problems that had produced successive delays. Finance was now
secure, and a steady, if small, revenue was coming from the
section that was operating. The Government funds had been
sufficient to see the cut made and the two seas joined, with the
prospect of a high income from tolls that, it was hoped, would
justify those who had campaigned for the construction over 30
years earlier. The worst of the structural problems had been
overcome, and while the cut would still need to pass through
whinstone and red sandstone, the contractors had gathered
expertise over the previous years that would make the problem
less serious. Furthermore, the opening of the Kelvin Aqueduct in
1787 gave a publicity boost to the Company that justified the
claim that the work was of national importance. Designed by
Robert Whitworth and constructed between 1787 and 1790, it
was at 400 feet long and 70 feet high the largest structure of its
kind and, with arched spandrels on the buttresses supporting its
four arches, it had cost £8500 to build (and was £2300 over
budget) but was regarded with great pride by Glaswegians. If
further funds had been required then, the public would not have
hesitated to give it support.

Control from London, resented as it was by both Glasgow and
Edinburgh, gave the Company a new pattern of organisation that
was more efficient and 'modern' than that which had caused
trouble in the past. When a hogshead of water from the Forth
was emptied into the Clyde on 28th July 1790, by a delegation of
Glasgow Magistrates and members of the Committee of Management of the Canal Company, it marked not only the opening
of the new Canal and the end of a 32-year saga, but the
introduction of a new system of large-scale organisation that

others were to copy in Scotland and elsewhere. The following month, the fisheries sloop *Agnes* left Leith and sailed to Greenock through the Canal, becoming the first vessel to complete the journey from East to West Sea in that manner.

8

The Canal in Operation

The chief advantage of canals was the easy transport of bulk cargo without hold-ups by rain, snow or fog. They were also much faster than carts and much cheaper as they did not demand tolls, as the roads did. This had been a reason for the involvement of the merchant classes in the promotion of the Forth and Clyde. Shareholders had accepted that profits would not be forthcoming for some time, but were also anxious that the Canal should start to pay its way as soon as possible. Consequently, when the cut reached Kirkintilloch in 1773, the Canal was opened to allow vessels to travel between there and Grangemouth. As a result, Kirkintilloch became a pivotal point on the waterway, with goods being shipped to and from Glasgow. Success there prompted merchants in the city to raise the funds to extend the Canal and bring it into Glasgow: to Stockingfield in 1775 and Hamiltonhill the following year. By 1776, the Canal had achieved the principal aim of the Glasgow interest by linking the city with the east coast.

By allowing the free movement of goods from the east coast, the Canal encouraged the development of the City of Glasgow, and this growth prompted an even greater demand for goods and materials transported by water. When the Canal was eventually completed in 1790 and links to Bowling and Port Dundas opened, the Forth and Clyde had accomplished its three main objectives: coast-to-coast travel for fishing and coastal vessels, inland communication for local manufactures, and a highway linking Glasgow and its hinterland with Edinburgh, Perth and Stirling.

In addition to through-going vessels, the Canal was used for local traffic, and proprietors along its banks, like the Carron Company, used their own boats. Farmers made use of small scows and lighters to move fodder, equipment and manure to their fields, and from 1800 the Carron Company ran daily 'market boats' between Grangemouth and Glasgow to allow them to take fresh produce to the city. In 1830 the Canal Company introduced 'cart boats' onto which farmers could run their carts from the banks of the Canal. These early roll-on-roll-off ferries saved a great deal of time in the loading and unloading of goods, and were extremely popular.

Work had started on the Monkland Canal in 1770 in order to bring coal from the rich Lanarkshire fields into Glasgow, where it was much in demand for domestic and industrial use. It was claimed that the waterway would halve the price of coal in the city by reducing transport costs. The 90-feet-high hill at Black-hill delayed the completion of a link with the Forth and Clyde, as had been intended, and the coal had to be loaded into boxes and hauled over the hill to continue its journey to Port Dundas: a process that was slow, costly and often made more difficult by snow or rain.[99] William Stirling, who had an interest in both canals, took over the Monkland and had a set of locks constructed at Blackhill to link the two parts of the Canal and complete the junction with the Forth and Clyde at 'Cut of Junction' at Port Dundas. As a result shares in the Monkland rose dramatically: where they had been £125 in 1770 and fallen to £25 in 1781/2, they soared to £1550 in 1793 when the junction was effected. As predicted, the price of coal in Glasgow fell and it became one of the principal cargoes on the waterway, with over 27,000 tons being carried in 1800, rising to 47,000 tons in 1804, and reaching over 80,000 tons in 1808.

The success of the Forth and Clyde in making coal available over a wide area was one of the stimuli to the opening of the Edinburgh Union Canal. Magistrates in the capital sought an Act of Parliament to create their own canal to the Monklands, but argument over the route made progress extremely slow. Between 1791 and 1817, debate raged over whether to compete with the Forth and Clyde Canal or to use it for at least part of the route, colliery owners arguing in favour of a line that passed by their mines, while landowners tried to benefit by selling land at inflated prices. The impasse was resolved by the Resident Engineer, Hugh Baird, proposing that a link be made with the Forth and Clyde at Lock 16, which would save money, and offering Edinburgh 'an ample supply of water . . . without any expense whatever from the community'.[100] This offer won over the magistrates, and in 1817 an Act was passed to allow the construction of a five-feet deep canal between Edinburgh and Falkirk at Lock 16. With the difference in depth between the two canals, through passage would not be possible for all vessels.

The Forth and Clyde, like the Union and the Monkland, had been built to allow the movement of goods that were bulky or heavy. Glasgow and the surrounding districts were able to obtain grain from North-East Scotland and the Baltic, timber from Scandinavia, manufactured goods from North-East England and wine from the Continent. In return, tobacco, sugar, textiles, iron ore and, especially, coal passed from west to east along the waterway. Charges were levied on the basis of distance and weight, typical costs being between ½ d. and 3d. per ton per mile, with coal and timber, essential for the shipbuilding and engineering industries, carrying the higher rates.

However, from an early stage it had been recognised that the Canal offered a suitable means of passenger transportation. The boats offered a much more comfortable journey than coaches, and the provision of reading materials and games for entertainment, toilet facilities and heated cabins in adverse weather made them attractive to the travelling public. In 1809 two passage boats, the *Charlotte* and the *Margaret*, carried passengers only, and operated from Port Dundas in Glasgow to Lock 16 at Falkirk, covering the route in five-and-a-half hours. By August 1812 this had become a regular service, the fare being 1/- per six-mile stage first class and 8d. second-class cabin, and the following year there were four passenger boats operating each weekday.

Almost from the beginning of passenger services, the Canal was integrated with the coach service to provide access to most of central Scotland, and by 1831 the minister at Kirkintilloch could report in the New Statistical Account that 23,170 people were using the station there. From Castlecary, and later Wyndford Lock, coaches conveyed passengers to Stirling, Dunblane, Crieff and Perth, while from Port Downie at Lock 16 passengers could travel on to Edinburgh, Alloa and Kirkcaldy. When the Edinburgh Union Canal was opened, they transferred from the Forth and Clyde to the fleet of boats operating on the shallower waterway. This network soon extended to provide connections with ships sailing to Liverpool and Ireland from the west coast and the Low Countries, Hamburg and the Baltic from the east. Such was the importance of passenger traffic that passage boats were given priority over other vessels, which were required to drop their tow lines to allow them to pass, or risk having them cut, quite legally, by the scythe-like prow of the 'Swift'. Night services were begun between Glasgow and Edinburgh and the boats had lanterns that shone light onto the towpath and water ahead to allow the horsemen to see ahead.

In 1830 the Forth and Clyde introduced a new type of vessel, the aforementioned 'Swift'. Based on the experiments of Thomas Neilson on the Monkland Canal, these boats were made of iron and were long and narrow; pulled by two horses with a postilion mounted on the second, they could reach speeds of nine or ten miles per hour. They reduced the time between Glasgow and Falkirk to three hours, and one vessel, the *Zephyr*, halved the journey time to two and three-quarter hours.[101] The public was impressed with the speed, and they became extremely popular. The Canal Company also gave these 'Swifts' special status by having the staff wear distinctive uniforms and the captain gold braid. The same vessels were used on the night service and were known as 'Hoolets', from the distinctive sound of their horn, which was used to warn other vessels of their presence. At the peak of the Canal's popularity, five boats left Edinburgh every weekday and three during the night. Competition from the emerging railways obliged the Company to reduce its fares after 1842:

1822:   Edinburgh to Falkirk   price 3/9d.
        Falkirk to Glasgow   price 3/3d.

1842:   Edinburgh to Glasgow   price 3/6d. (First Cabin)
                                        2/4d. (Second Cabin)
1844:   Edinburgh to Falkirk   price 1/4d.
        Falkirk to Glasgow   price 1/2d.

The 'Swifts' built up a reputation that became part of the legend of the Canal, but despite their popularity they could not beat off the competition from cheap and speedy rail travel.

While ships could use sails if the wind was favourable, they normally had to be towed by horses. This was the source of much delay to through traffic as vessels had to wait for horses at the sea locks to allow them to continue their journey. The Canal Company was anxious to reduce delays resulting from this congestion and was impressed by the twin-hulled paddle steamer *Experiment* that William Symington had used on the Monkland Canal in 1789, and commissioned him to experiment with steam tugs. In 1802 he produced the *Charlotte Dundas* and demonstrated that it was possible to use steam power to haul the barges through the Canal and reduce the dependence on horses. The Committee of Management considered using the vessel, but feared that the wash from it might damage the puddle lining the canal banks and decided to ban the use of steam power on the Canal. As a result, horses continued to operate for another 20 years, and in some places into the next century, rendering the

Canal unable to compete with the railways when their challenge emerged later in the century.

Thomas Grahame, a member of the Canal Company Council, hired a small paddle steamer from David Napier in 1825, and demonstrated that it could tow passage boats at a steady five miles per hour more efficiently, and, very importantly, more cheaply than horses. He also demonstrated that the wash did no damage to the canal banks, and the Company lifted the ban on steam boats and placed orders for paddlers for use on the waterway. Despite this, horses were still preferred to haul the small lighters that operated as freight carriers, and the paddlers were to be used for the passenger vessels. With the threat from the railways becoming obvious in 1840, the Company encouraged experiments with screw propulsion. In 1845 the *Firefly* was developed for canal use, but its introduction came too late to save the Canal, as passengers were already moving to rail travel:

> Had she come into the canal in the previous decade, or had the 'Swifts' not succumbed to competition from the railways, she might have provided the answer to economical towing with speed.
>
> (Paul Carter, *Forth and Clyde Canal Guidebook* 1991, page 39)

Attempts to introduce steam power continued over the next 10 years. In 1856 James Milne, the Resident Engineer at Hamiltonhill, converted the lighter *Thomas* to produce a steam-powered screw craft that proved highly efficient. The design was taken up by Swan Brothers at Maryhill, who built the *Glasgow*, the first purpose-built screw lighter for use on the Canal. To aid draughting the engine, the exhaust was turned up through the funnel and the puffing noise that this produced led to such vessels being called 'Puffers'. Several were built for use on the canals, where they were referred to as 'inside boats' to differentiate them from the larger versions, 'outside boats', that several owners, like Carron, Salvesen, Hay and Burrell, had built for coastal waters and which were to earn fame as the mainstay of goods traffic along the coast and into the islands well into the twentieth century.

It had been anticipated that the Canal would stimulate industry along its length, but even the most optimistic of the promoters could not have foreseen just how important it would become. The need for vessels on the waterway meant that shipbuilding became one of the first industries to set up beside it. James Welsh had been building sloops and lighters at the eastern sea locks from 1789, and others followed his lead. Within a few years of the opening of the waterway, shipbuilding yards, complete with carpenter shops and graving docks, were to be found along the Canal, from Grangemouth to Falkirk and through Kirkintilloch, Hamiltonhill and Maryhill to Bowling. The timber for construction was supplied, naturally enough, through the Canal itself, but after 1830 the majority of the vessels were of iron construction and the yards had the sheets brought by water. When the railway network expanded, some of these yards closed down as demand fell, and others relocated to coastal areas which could be supplied by rail. Kelvin Dock, however, remained in operation until 1949.[102]

As the Canal passed through an area rich in mineral resources and provided a convenient means of transporting raw materials as well as manufactured goods, it became very attractive to industrialists. With a plentiful supply of water for steam engines, it was a natural location for iron foundries and, later, engineering shops. Bairds was established at Hamiltonhill before 1800 and the iron industry spread along the Canal in the following years: John Neilson at Oakbank, Burnbank Foundary at Bainsford, Taylor's Engineering at Falkirk, Port Downie Iron Works at Camelon, Smith and Wellstead at Bonnybridge, South Bank, Lion and Basin Foundries at Kirkintilloch, Lochburn Iron Works at Possil, Lambhill Forge and Foundry, Firhill Iron Works and Caledonia Foundary at Firhill, Maryhill Iron Works, and Yuill and Wilkie at Springbank. These industries provided employment that attracted workers into the area and this led to massive housebuilding projects alongside the works. Thus small villages became towns, and this in its turn attracted other businesses, shops and services to meet the needs of the expanding industrial population. At Glasgow and Falkirk, distinct and separate villages expanded and merged into larger towns. In Glasgow the population, which had been 28,000 in 1761 and had risen to 65,000 in 1790, reached 110,000 in 1811 and continued to grow to reach almost 1,000,000 by the end of the century. The minister at Kirkintilloch reported in the Statistical Account that the population there had doubled since the opening of the Canal and the provision of employment:

1791 population 2,640
1801 population 3,210
1831 population 5,888.[103]

Foundry workers lost fluid as a result of the high temperatures in which they toiled and were given beer at the end of the shift to replace it. This created a demand for beer and other alcoholic beverages, and distilleries and breweries saw the advantage of setting up close to the foundries and beside the Canal which could transport the grain to their very doorstep. This accounts for the appearance of Rosebank Distillery at Falkirk and Bankier at Wyndford, while Port Dundas claimed to have the largest distillery in the world.[104] Most constructed their own loading places on the Canal to unload the cargoes of grain from North-East Scotland and the Baltic. Such was the demand for grain that it became one of the principal cargoes carried on the Canal and the one that produced the highest revenues, bringing in an average of 20% of all canal revenue, and forming 31% in 1831

(see Appendix C). One of the effects of this was to bring down
the price of grain:

> The trade upon it is already great and is rapidly increasing.
> One of its first effects has been, to equalize, in great
> measure, the price of grain, throughout all the corn-
> counties in Scotland: to the temporary loss of the land-
> holders, in the southern, and to the gain of those in the
> northern districts.
>
> (Rev. William Dunn, Old Statistical Account,
> Vol. II, Dumbarton, pp. 77–8.)

Coal and timber were the other bulk cargoes that used the Canal
and produced significant income. Although both fluctuated from
year to year, coal consistently accounted for an average of 9%,
and timber about the same proportion, of revenue. Loading
places, called coupes, were built along the banks of the waterway
to permit barges to take on coal, and Baird's colliery at Twechar
and Gardner's at Kirkintilloch each had their own coupes. The
timber basins at Grangemouth, Kirkintilloch, Firhill and Port
Dundas, with smaller 'ponds' at Maryhill, Temple and Bowling,
were used to store supplies from the Baltic and Scandinavia.
Around these basins there developed sawmills which made use
of the Canal to supply water for their steam engines and which
provided further employment on the Canal. Sawmillhill Street at
Port Dundas is a reminder of the industry once located there.

Glasgow saw the development of the glass industry as the
sand deposits in the Kelvin Valley came to be exploited. Victoria
Glass Works was established at Hamiltonhill, the Glasgow Glass
Works nearby and a Glass Bottle Works and the Forth Glass
Works both at Firhill. Industrialisation in the city also saw the
emergence of chemical manufacturers at Kelvin Dock, Port
Dundas and Firhill, and a Paraffin Works at Stockingfield.
Outside the city boundaries there were similar developments

as a result of the ease of transporting raw materials from different locations to the works and thence to all parts of the kingdom. Ross's Chemical Works and Scottish Tar Distillers opened at Camelon and Falkirk, and the New Caledonia Mines Company built a nickel smelter at Kirkintilloch.

Without the Canal, it is difficult to imagine the growth of industry in the central belt. It is doubtful if the Lanarkshire coal deposits could have been successfully exploited without the facility to transport the coal to Glasgow and from there to the rest of Scotland and beyond. Similarly, the iron industry at Gartsherrie, Cleland, Dundyvan and Clyde works would not have existed without the Canal. At best these would have been small local firms supplying a limited, largely local market. The cargoes of stone from the quarries at Twechar and Auchinstarry had to be carried to the towns, and Glasgow in particular, to meet the housing needs of an expanding population. The ability to move raw materials and the products of these industries, in bulk and cheaply, was essential to their growth and the development of an industrial society in Scotland.

There were others for whom the Canal offered an entrepreneurial opportunity. The need for cadavers to supply the Anatomy Department of Edinburgh University encouraged some 'Resurrectionists' to convey the bodies of recently buried people from Cadder by canal. In the 1820s it was discovered that a number of casks labelled 'Bitter Salts' being transported by barge to an Edinburgh address actually contained human bodies destined for the University. Perhaps it was these events that encouraged two of the Canal's most famous labourers, William Hare and William Burke, to diversify and supply even fresher bodies than those from graveyards.

9

The Decline of the Canal

On 6 June 1831 George Stephenson was asked, by a number of people who had been impressed by the success of the Liverpool and Manchester Railway, to advise on the construction of a railway link between Edinburgh and Glasgow. He suggested a connection from Edinburgh to Coatbridge to join the Glasgow and Garnkirk Railway that Thomas Grainger and John Miller had proposed. The Forth and Clyde Canal Company quickly realised the danger that this would pose to its virtual monopoly and vigorously opposed the plan. During the debate in the House of Commons in May 1832, the canal interest laid emphasis on the differences between these proposals and the Liverpool and Manchester scheme, and the Bill was rejected.[105]

However, the merchants in Edinburgh and Glasgow were attracted to the notion of cheap transport for both goods and passengers and decided to pursue matters further. It was decided to raise another Bill to have the railway built, and finance was attracted from England. Further study revealed that the best route lay in the 'valley along which the Forth and Clyde Canal passes'. It was realised that the Canal Company would again oppose the Bill, and to avert at least some of the complaints, the proposers stressed that the railway would carry only passengers and that freight should remain on the Canal. It is doubtful if this tactic would have succeeded, but the railway interest was to benefit from a lack of unity among the canal companies. The Edinburgh Union saw an opportunity to compete with the Forth and Clyde by means of a rail link from Glasgow to Falkirk, bypassing the Great Canal, and in 1835 helped finance the Slamannan Railway Act to further this aim. Consequently, when the Edinburgh and Glasgow Railway Bill was presented to the Commons in 1838, the canal interest lost the argument and on 4 July an Act was passed authorising the building of the railway.

Although defeated on the Bill, the Forth and Clyde Canal Company was not prepared to docilely accept the coming of the railway and was prepared to fight its corner and use obstructive tactics where it could. Thus, when plans were put forward to build a bridge to take the railway from Queen Street, in Glasgow, over the Forth and Clyde at its junction with the Monkland Canal, the Company raised objections. As a result, the Edinburgh and Glasgow Railway was obliged to dig a 1,000-yard tunnel under the 'Cut of Junction'. In addition to the added expense of tunneling, the descent was so steep that the Company employed a static engine to haul the trains up into the station, and later required two engines for that purpose.[106]

The opening of the railway in 1842 provided a cheap service between Glasgow and the Capital that was much faster than the Canal and attracted passengers from the Forth and Clyde. The Company, which prior to 1842 had been carrying up to 197,000 passengers a year, found that the numbers dropped rapidly. When the Caledonian Railway Company began operating services through Castlecary in 1848, it brought an end to passenger traffic on the Canal. On 11 March 1848 the Forth and Clyde announced the end of its passenger services. A. & J. Taylor, however, continued to operate 'Swifts' between Port Dundas and Wyndford Lock to provide a service to the villages along the route. George Aitken of Kirkintilloch took over from the Taylors and operated extremely popular pleasure trips to Craigmarloch until 1880.

The Edinburgh and Glasgow Railway had not, at first, sought to carry coal, partly to appease the Canal Company and partly because the need for level ground had taken the route past the main coal-producing areas. Rich coal deposits were discovered close to the rail line at Polmont in 1842 and the Redding

Colliery decided to transport its coal by rail in October of that year. As the rail network expanded, providing access to greater markets, other collieries followed the lead of Redding, and within a decade most of them had links to the Edinburgh and Glasgow Railway, depriving the Canal of another of its sources of revenue. One exception was the colliery at Twechar which had agreed to convey coal by canal in return for being provided with a rail swing bridge and which honoured the agreement long after others had ceased to use the Canal.[107] Even the Carron Company, which had been involved in the earliest proposals for the Canal, now turned to the railways. There was little that the Canal Company could do to reverse the trend as its freight traffic went the way of its passengers.

In 1845 proposals were floated to amalgamate the Edinburgh and Glasgow, Scottish Central and several Lanarkshire mineral railways with the Forth and Clyde, Monkland and Edinburgh Union canals. This caused alarm among the merchants of Glasgow and Edinburgh, who feared that this would create a monopoly of transport in central Scotland that would be able to maintain high prices. They began a compaign of protest, claiming that any amalgamation would be ruinous to trade. English investors in the Edinburgh and Glasgow Railway believed that the Scottish businessmen were not concerned about capitalising on their transport monopoly and were more interested in keeping transport costs low to promote their other interests. As evidence, they cited the case of John Wilson of Dundyvan. Wilson was a director of the Edinburgh and Glasgow Railway Company, and also a leading campaigner on behalf of local traders who were agitating for lower transport rates. The Select Committee set up to examine the amalgamation proposals announced, on 1 July 1846, that it was unable to agree on the type of amalgamation required as the rail network was not sufficiently developed at that time. In September the English shareholders forced the resignation of the Board of the Edinburgh and Glasgow Railway and replaced them with Scots members who were more sympathetic to the amalgamation.[108] A sustained campaign was now mounted to win support for the amalgamation of transport services in the central belt, with the railway as the senior partner.

In May 1848 the Scottish Central Railway linked with the Edinburgh and Glasgow at Greenhill, opening connections to Perth and Aberdeen in the north and London in the south. 'Railway Mania' was at its height, and within three years the rail network covered most of Great Britain and opened the entire country to rail traffic. This made way for the amalgamation proposals to be put forward for consideration again.

In the twelve years after the end of passenger traffic, the Forth and Clyde was still profitable. Income varied between £44,000 and £52,000 a year, and expenditure averaged around £15,000, while dividends were between 7% and 8%. In 1845 the Canal Company bought the Edinburgh Union Canal, which had proved unprofitable and had virtually closed. But four years later the amalgamation that had been proposed earlier took place, at least in part, when the Edinburgh and Glasgow Railway and the renamed Forth and Clyde Navigation Company merged. With the rail interest now firmly in charge, this was to lead to the eventual closure of the Canal.

It was a suggestion from the Canal Company that led to the creation of a rail link from Grangemouth via the Stirlingshire Midland Junction in 1860, which was intended only for freight, but which started carrying passengers the following year. This meant that the Caledonian Railway, when it acquired the Port of Grangemouth in 1867, was obliged, as a condition of the sale, to purchase the Forth and Clyde Canal.

There was concern among canal companies that they might be taken over by railway companies, whose intention was to close them down. This had apparently been the fate of the Aberdeenshire Canal which the Great North of Scotland Railway Company purchased in 1845 and closed to commercial traffic in 1854.[109] To secure the future of the canals, the Government passed an Act requiring the railway companies to maintain and keep navigable all waterways that they managed:

> Every railway company owning or having management of any canal . . . shall at all times keep such a canal . . . thoroughly repaired and dredged and in good working condition so that the whole . . . may be at all times kept open and navigable for the use of all persons desirous to use and navigate the same without any unnecessary hindrance, interruption or delay.
>
> (Regulation of the Railway Act, 1873)

This Act, together with the efforts of the Forth and Clyde Navigation Company to attract business, helped keep the Canal in operation for almost another century. Between 1893 and 1939, James Aitken ran pleasure cruises from Port Dundas to Craigmarloch, and occasionally as far as Lock 16. These trips were extremely popular and allowed the public to escape the industrial noise and pollution of the city and enjoy a day out in the countryside. As well as providing entertainment on the steamers – in particular the famous *Fairy Queen* – Aitken provided an inn and picnic area at Craigmarloch, where families could relax:

> On Saturdays during summer months the steamship *Gipsy Queen* sailed from Port Dundas in Glasgow with happy passengers for a trip along the canal to Craigmarloch Bay. There the passengers disembarked for a couple of hours where they could lunch in the large café or picnic on the hillside. There were plenty of swings for the children and a good time was had by all.[110]
>
> (Hugh Ross of Kilsyth, *Memories of a Canal Trip*, 1913)

The *Gipsy Queen* operated up to the start of the Second World War, when her service was discontinued.

Despite the lasting legacy in popular memory of the pleasure steamers, the Canal still functioned primarily as a means of hauling freight. Up to the outbreak of war in 1914, the Forth and Clyde continued to bring in an income of between £35,000 and £87,000 a year and was running at a profit. In 1923 it became part of the London and North-Eastern Railway Company, which was empowered to close the waterway from Lock 16 to Port Downie. This it did, and the flight of locks that had linked the Union and Forth and Clyde was removed, although the Union was allowed to remain open to supply water to the industries along its banks.

The closure to commercial traffic of the sea locks at Grangemouth during the Great War was a significant factor in bringing about the demise of the Canal, although there was considerable movement of coal and timber within the canal area due to the increased demand during the conflict, over 85,000 tons of timber and 182,000 tons of coal passing in 1913 alone. Pig-iron and other commodities ceased to be carried through the Canal from sea to sea and, with the end of the war, the trade did not recover. Several writers, among them David Bolton and Paul Carter, have suggested that the closure of the waterway would not have been entirely unwelcome to the rail companies that owned them and regarded them as an alternative to their monopoly of transport. What is obvious, however, is that the Forth and Clyde, like its counterparts in England, did not receive the funds required to ensure adequate maintenance, especially during the slump of the 1920s and the depression of the '30s.

As a result of the Government's nationalisation programme in 1948, the waterways were put under the jurisdiction of the British Transport Commission, and this produced a new set of difficulties. Competition from road transport was applying pressure to the railways, and a new lobby emerged that urged the advantages of the road system and argued for the creation of a network of motorways on the continental model. The newly privatised road haulage industry was in the forefront in pushing for improvement of the road system, which would benefit their business. Some enthusiasts even proposed that filled-in canals could be built on fairly cheaply, and because of the directness of their routes, would make ideal highways for motor vehicles. With a struggle for survival going on between the railways and the roads, the canals were regarded as expendable by both groups.

In 1955 the Board of Survey, set up by the Transport Commission, issued its report on the condition of the inland waterway network. The authors divided the waterways into three distinct groups. Group 1, some 336 miles, was considered suitable for development; Group 2, around 994 miles, they considered worthy of being retained as having some potential; Group 3, the remaining 771 miles, and including the Forth and Clyde and the Edinburgh Union, was deemed as 'having insufficient commercial prospects to justify [its] retention for navigation'.

This decision is surprising when the facts are examined. Far from being derelict, the Forth and Clyde was in regular use, mainly by fishing vessels passing through the safe waters of the Canal to move from the herring to the white fish grounds each year. Ironically, the Canal had just achieved one of the aims of the original scheme put forward 290 years before, when a warship, the midget submarine XE1X, passed through the waterway.[111] In 1956 the canal was used by 189 pleasure boats, 98 fishing vessels and 14 cargo ships.[112] Admittedly, income from these vessels was not nearly enough to meet the cost of maintenance. In 1955 income was £29,545 and expenditure was £65,286; and in 1956 income was £36,456 and expenditure was £99,291 (see Appendix D). However, this cannot be considered particularly great when compared to the working deficit of other Transport Commission enterprises. The decision to close the Canal seems to have had more to do with the financial needs of the road lobby, which had objected to the cost of renewing 50 bridges that crossed the Canal. Work on the Denny bypass on the A80, it was argued, would be much less expensive and end the bottleneck caused by the old swing bridge if a fixed bridge was built over the Canal rather than installing a lifting bridge to allow craft to pass through. Parliament decided that road transport had priority and authorised the closure of the Canal rather than find the £160,000 needed to replace the swing bridge:

> It was a suspiciously convenient reason to close a canal that had become politically unpopular.
>   (Guthrie Hutton, *Forth & Clyde Canalbum*,
>   Ochiltree, 1995, p. 24)

Those interested in keeping the Canal open warned of the effect that closure would have on the fishing industry. The Bowes Committee had stated that preventing vessels from passing through the Canal would have a 'severe impact' on fishing. In the middle of the century, fishing boats continued to use the Canal, but in diminishing numbers: 189 in 1953, 153 in 1954, 119 in 1955 and 98 in 1956. Despite appeals by fishermen, the Government decided that the interest of road transport was paramount:

> . . . having given the fishermen a full opportunity to state their case, I have concluded that the balance of advantage lies in saving public money on road bridges rather than continuing . . . a 'disguised subsidy' to a small section of the Scottish fishing industry.
> (Lord Craigton, Minister for Roads and Fisheries, 1960)

The rapid expansion of road transport, particularly the haulage sector, had attracted traffic from the railways, some lines had become uneconomic, and the Government tried to address the problem by setting up a committee to examine the whole issue of transportation. The committee's report resulted in the Transport Act of 1962, which ended the British Transport Commission's control of the system. Although principally concerned with the difficulties of the railways and laying the foundation for Dr Richard Beeching's programme of modernisation, the Act created a separate waterways authority. As David Bolton, in *Race Against Time*, remarked, the waterways that had been bought up by their rail rivals, and then nationalised under railway control, were now free to pursue their own policy. However, their freedom of action was severely curtailed as the Transport Act brought an end to the protection that they had enjoyed under the 1873 Act. Clause 43 of the new Act allowed the British Waterways Board to make changes to their services and facilities 'as they think fit', and permitted the sale of property and even the canals themselves.

On 12 and 13 October 1962, the MV *Ashton* became the last passenger vessel to use the Canal, travelling the round trip from Bowling to Kirkintilloch. In November, a few fishing vessels passed through for the last time. In January 1963 the Forth and Clyde Canal (Extinguishing of Rights of Navigation) Act came into effect, ending the original purpose of the Canal and authorising the British Waterways Board to dispose of its assets. There was no public outcry at this as the Canal had become unpopular and a number of drownings had led to demands that it be filled and put to other use, and this campaign, aided by the road lobby, was given wide coverage in the press. Politicians were quick to assure the public that the 'Killer Canal' would be filled as soon as the money to do so became available. In the year that the Canal ceased operating, the road interest pushed ahead with its plans for a road bridge over the A80 near Castlecary, effectively blocking east-west navigation. Outside the urban areas, however, the Canal continued to be used for recreational purposes. Anglers had long made use of it and were not unduly perturbed by its closure, and sailing in small boats never really ended. In some rural areas, it would be true to say that the ending of navigation was not noticed to any great extent as the volume of traffic had been in decline for many years and the main use was local, and remained so.

'The old canal may be dead, but it won't lie down.'
(Paul Carter, *Forth & Clyde Guidebook*, 1991, p. 47)

10

Revival

A number of factors converged in the early 1970s to revive interest in the Canal. The industrial decline that had marked the post-war years continued throughout the 1960s and there was a shift from the traditional heavy industries of shipbuilding, coal and iron to new industries employing fewer people and the consequent relocation of the population in the growing housing estates on the outskirts of the towns. As the developers moved in, there was a risk that the industrial past could be lost forever, obliterated or buried under the new buildings. This led some historians to attempt to preserve, or at least record prior to their destruction, the monuments to Scotland's industrial past. By 1966, groups of industrial archaeologists, many of them students of John Butt and John Hume of Strathclyde University, were undertaking what could be termed 'rescue archaeology' by moving onto sites to record them before they were razed by the bulldozers. The Forth and Clyde Canal fell into the category of such monuments and attained a higher profile than before.

A second factor that operated in the Canal's favour was that it attracted interest as a community facility. The 20 years after the war had brought about much change in society in Scotland, with shorter working hours and longer holidays. While the Canal had always been popular with walkers and anglers, it now offered potential for further recreational use, with boating, model and full-size, being the most obvious. The realisation that the Canal could provide an amenity gave it a future and gradually changed the attitude of the public.

The third, and perhaps the most significant, factor was the formation of the Scottish Inland Waterways Association in May

1971. This organisation began to campaign for the restoration of all of the country's canals and, by raising awareness of their potential, encouraged other interested parties and local authorities to re-examine their attitude to canals in their own areas. It was quickly discovered that many parts of the Forth and Clyde had been used by local residents for the disposal of rubbish, and were not only unsightly as a result, but downright dangerous. Local volunteer groups took on the task of cleaning up the Canal, and this raised the profile of the waterway and gained support from the public. The tragic drowning of four children at Clydebank reopened the debate on the future of the Canal, and centred on whether a waterway that was in use would be safer than one that was derelict: infilling does not seem to have been proposed as a strategy.

The potential of the Forth and Clyde for recreational use was first raised at semi-official level by the landscape architects William Gillespie and Partners in 1974. It was suggested that the waterway provided an opportunity for integrated leisure amenities in urban areas, and local authorities took up the idea and examined its application in their own patch. The Strathkelvin Canal Park is one example of this new and enlightened approach to the Canal. Within a few years, the area around Kirkintilloch witnessed regular boat and barge trips.

Most of the early clean-up campaigns, which had no official status, could be ignored by local authorities and were by some, who regarded the campaigners as enthusiastic 'do-gooders'. However, these volunteer groups attracted publicity, and this led to their being given the support and often sponsorship of local organisations. When Strathhclyde Regional Council was created in 1975, it discovered that 25 of the Forth and Clyde's 31 miles lay within the regional boundaries, and determined to make use of it. The Council's Leisure and Recreation Department was given the responsibility of ensuring that the Canal was available for use by those who were interested in leisure pursuits.

The Department went further than this and produced a series of leaflets on various aspects of the Canal, its history, its wildlife and the memories of people who had used it. The support of the Regional Council meant that progress was possible, as it would listen to people with ideas for using the Canal, and had funds that could be used to put the plans into effect.

The most effective method of promoting the Canal was to raise public awareness by having the waterway used. Boat rallies were held at several places on the summit to encourage local communities to become involved in making use of the facility. The biggest boost came when Radio Clyde announced its intention to back the Canal as its Community Involvement Scheme for 1976. This brought together canal enthusiasts and members of the local community groups in efforts to organise events at different locations on the Forth and Clyde. The impact was such that local authorities, at District and Community Council levels, decided to take part in the event and afterwards continued their support. By the end of the decade, many of the authorities were making use of job creation schemes to help with the cleaning up of the Canal and improving the environment surrounding it.

The formation of the Forth and Clyde Canal Society in 1980 marked a turning point. Now there was a group of enthusiasts prepared to devote time to advancing the cause of restoration. It worked alongside the British Waterways Board and local bodies to promote the Canal, and arranged exhibitions and meetings which raised public awareness. When the Forth and Clyde Local Plan was proposed, the Society produced evidence to support reopening the Canal to navigation and submitted recommend-ations to Strathclyde Regional Council which had declared an interest in the scheme. Progress was slow, but advances were made and the Council was persuaded to make funds available for restoration projects. Together with British Waterways, plans were made and implemented to improve the waterway, rebuild bridges and restore buildings associated with it. The revival had begun.

As the Canal Society was beginning its work, it had a piece of good fortune that enabled it to put into effect its aim of having the Canal brought back into use. In 1980 the Clyde Port Authority decided to retire its three remaining cross-harbour ferries and the Society's offer to purchase them was accepted. The press, which 20 years earlier had made references to 'The Filth and Slime Canal' and the 'Canal of Death', now positively enthused over the prospect of the Govan Ferry sailing on the waterway. Both the *Glasgow Herald* and the *Scotsman*, in a rare display of Edinburgh and Glasgow togetherness, praised the Society for bringing the No. 8 Ferry to Kirkintilloch (perhaps ironically it was transported there by road!), when the vessel, renamed *Ferry Queen*, was launched on the Canal in April 1982. Ferry No. 10 became the *Caledonian* and operated as a restaurant boat from the Stables at Kirkintilloch, travelling as far as Craigmarloch, together with its companion the *Lady Margaret*. Within a decade there were several vessels on the Canal, stationed at various locations between Glasgow and Falkirk.

British Waterways announced, in 1988, the Glasgow Canal Project, which aimed to re-open the Forth and Clyde Canal to navigation on a commercial basis. The growth of the leisure industry over the period, together with the success of the craft operating on the Canal, led them to believe that such a scheme was viable. The original intention was to reopen the waterway between Temple and Port Dundas on the Glasgow branch and then on to Kirkintilloch on the main branch. It was believed that this would make the canalside more attractive to property developers and encourage housing and leisure facilities, and thus increase the value of the land.

The task facing the developers was considerable. During the years of inactivity, weeds had taken over much of the canal, growing on the deposits of silt that had accumulated at the bottom. Further, some of the locks were in need of major repair, water seeping through them and reducing the level in some places. Work began at Maryhill, where the locks were repaired and the Canal cleared of weeds, and the section opened on 25 June 1988. To celebrate the opening of this first section, the *Ferry Queen* was hired to take a party of guests on a cruise. Unfortunately, the vessel became embedded in the mud and silt and only reached Maryhill thanks to the efforts of Strathclyde Police, whose officers hauled it along on ropes. It was, perhaps, fortuitous that their Divisional Commander chanced to be one of the VIPs aboard.[113] The result was that the decision was taken to dredge the Canal to make it sufficiently deep to take fully laden vessels.

By a happy coincidence, the 200th anniversary of the opening of the Canal was celebrated the same year that Glasgow was to be the Cultural Capital of Europe. This gave the waterway a higher profile, and several events were planned to attract the public. Over 7000 people visited the Canal in Glasgow, and a two-day rally at Maryhill toward the end of the year attracted a crowd of almost 20,000. Falkirk was chosen as the venue for the Inland Waterways Association's National Trailboat Rally, and this brought representatives from every canal society in the country to the Forth and Clyde. To mark the bicentenary, a fleet of small craft carried water from the Clyde and poured it into the Forth: a re-enactment in reverse of the opening ceremony in 1790.

The success of the 1990 events led enthusiasts to put forward plans to make the whole of the Canal navigable from the Clyde to the Forth. Local authorities implemented a policy of resurfacing and replacing towpaths to open the Canal to the public, it being argued that this would reduce the vandalism that had been a problem over the years and which was costing large sums of money.

There were considerable obstacles to the achievement of the plan. Three sections of the Canal had been infilled at Grangemouth and Clydebank, including two of the locks, but the total lost was less than one-and-a-half miles. In contrast to the Monkland Canal, which had been built over by the M8 motorway, it was possible to re-open the channels on the Forth and Clyde. The most serious difficulty arose from the number of roads that had been built over the Canal, reducing the headroom in most cases and blocking it completely in others where it had been culverted, the A80/M80 being the greatest problem as it is the busiest road in Scotland. In addition, there were 12 bridges that would have to be replaced to allow the passage of vessels.

The greatest problem was, as always, finance. Local authorities had limited budgets and, however generous they were inclined to be, there was a limit to how much could be allocated to a non-priority area like the Canal at a time of demands from major areas like education and social work. By the mid-1990s a new source of money became available when the Millennium Commission was set up. This body was created to help celebrate the 'Year 2000' and was empowered to make funds available to bodies with schemes that would not only celebrate the Millennium, but also have a lasting effect on the community. In 1995 British Waterways, with the support of Strathclyde Regional Council and the Forth and Clyde Canal Society, submitted a bid to acquire funding for the waterway scheme. They proposed dredging the Canal to its original depth of eight feet, renovating or replacing all the locks, restoring the buildings alongside the Canal and raising the level of those roads that obstructed canal traffic. The scheme was very costly, and was rejected by the Commission as too ambitious.

Undeterred, the advocates of the Canal came together to produce a plan more acceptable to the Commission and to secure funding from other sources – an essential requirement for a successful bid. Costs were reduced by abandoning the proposed renovations at Port Dundas and agreeing to dredge the Canal to a depth of six instead of eight feet. By having a headroom of just under ten feet (three metres), the cost of reconstructing bridges and roads was also much reduced. It was also agreed that there should be a link with the Union Canal at Falkirk and that the long-vanished locks be replaced with a 'Wheel'. The plan attracted support from the local authorities in the east and west of Scotland, other public sector bodies, enterprise companies and the European Commission. The Forth and Clyde Canal Society also raised a petition containing 30,000 signatures in favour of the re-opening of the Canal. In 1997 the Millennium Commission accepted the new bid and made £32 million available toward the anticipated cost of £78 million, and the 'Millennium Link' was born.

Acceptance of the bid was only the start of the process of restoration. Local government reorganisation had produced seven councils that had to be consulted rather than the two Regional Councils as previously. The European Commission required to be supplied with full details before funds were released, and other participating agencies (see Appendix E) had to be consulted. Despite the logistics, British Waterways were able to win the active support of all the groups involved and work commenced in March 1999.

From the start it had been decided that the traditional design should be retained wherever possible and that even where the new structure was to be made of concrete, it should have a traditional appearance. Thus the replacing of timber locks and stone bridges called for the employment of carpenters and stonemasons on a scale not seen for many years. When the roads were being raised, it was discovered that the skills of the original builders had to be revived as modern techniques were not always satisfactory for the restoration work involved. The scale of the undertaking can be gauged from the fact that 24 bridges have been constructed to allow traffic to pass over the canal, two opening bridges have been reconstructed, nine new locks built and almost two miles of new channels dug, while 40 locks have been repaired and 34 pairs of new timber lock gates installed.

The central focus of the development was the construction of the Falkirk Wheel to connect the Forth and Clyde with the Edinburgh Union Canal. This structure, the first of its kind in the world, is a gigantic 115-feet-tall boat lift that takes eight boats at a time and transfers them between the canals in around 15 minutes. Led by British Waterways and utilising the experience of Morrison-Bachy-Soletanche, Ove Arup Consultants, Butterly Engineering and the architects RMJM, the Wheel and its associated structures cost some £17 million and employed over 500 construction staff. Although the Canal came into operation in May 2001, the climax of the Millennium Project was the official opening of the Wheel and its associated visitor centre by Her Majesty the Queen on 25 May 2002.

The stated aims of the Millennium Link were to improve the environment, restore sites along the banks of the Canal, support leisure and commercial developments and ensure access for all to the Canal. It was hoped that it would lead to increased recreational and tourist opportunities, provide employment potential and encourage commercial development. The restoration work began to bear fruit. On 12 July 2001, Scottish Enterprise Dunbartonshire announced plans, drawn up in cooperation with West Dunbartonshire Council, for the redevelopment of Clydebank on Clydeside, 'taking advantage of the canal as an environmental feature'.[114]

By 2002 it was again possible to travel by water from Bowling to Grangemouth and between Glasgow and Edinburgh. The original proponents of the Forth and Clyde must be smiling down on their Great Canal, which is once again fulfilling their dream.

Conclusion

A new phase in the history of Scotland was begun when Sir Lawrence Dundas cut the first sod on the banks of the River Forth on 10 June 1767, in the presence of a large body of workmen who were to start digging the Forth and Clyde Canal. After almost a century of surveys, discussions and eventual rejections, a navigation that was expected to usher in a new era of prosperity was about to be constructed. It was not foreseen that financial and other problems would result in another quarter-of-a-century passing before the work was completed and the East and West Seas at last joined, nor was it realised at that time that these very difficulties would help create a new Scotland, significantly different from the old one.

The failure to implement the proposed schemes in the seventeenth and early eighteenth centuries can be attributed to two main factors. In the first place, money on the scale required was beyond the means of the country. Scotland was too poor to finance such a venture before the middle of the eighteenth century. In addition, the confidence of any would-be investors had been shaken by the Darien disaster and failure of the Company of Scotland, and another generation would have to pass before Scots would be prepared to risk capital on a large-scale undertaking. The only source of funds of the magnitude needed was the Government, and this was beyond the means of the old Scots Parliament; and at Westminster, after 1707, there was little prospect of the Whigs, who dominated the House of Commons for the first half of the century, being persuaded to abandon their mercantilist principles and advance the money. As Defoe had predicted, capital did not become available until the wealth of Scotland had increased, and this took more than a generation after the Treaty of Union.

A second, equally important, factor was the will to undertake a work of such long-term duration. The Scots Parliament, or Estates, had been dominated by a landowning aristocracy which suffered enormous hardships during the severe agricultural depression of the 1690s which further eroded the losses many had suffered by investing in the Company of Scotland. When they considered trade, it was in terms of expanding the export of cattle to England, sending more coarse linen to Europe, or winning a share in the rich colonial trade dominated by the English. After 1707, these economic goals were largely achieved, but the Union left a large number of noblemen with a diminished role to play in the political life of the country, and they looked for other ways in which to exercise power. It has been suggested that an active role in commercial and economic enterprise became a substitute for political activity for this class. They took an active part in a number of societies, mostly based in Edinburgh, which sprang up during the century to encourage improvement, and which reflected the writings and philosophy of the 'literati' who advocated involvement in business life as suitable for members of a 'polite society'. While this helps explain the appearance of the will to invest in ventures, it could be argued that it did not represent such a radical change as has been suggested. A sense of duty had encouraged the gentry to participate in a number of schemes in the past, and the eighteenth century provided greater opportunities for such involvement.

Economic factors were, however, critical in initiating the moves to build the Canal. Most writers have stressed the rapid growth of Glasgow in the years after 1720 as a result of the tobacco trade, and it was this development that influenced the Scottish economy in a number of ways. It, firstly, created a class of wealthy merchants, the Tobacco Lords or Barons, whose influence on the political life of the city was considerable, and who produced a demand for luxury goods from the Continent that needed to pass through ports on the east coast, particularly Bo'ness. In the second place, the tobacco industry also created employment in the city and the labour force relied upon the rich agricultural lands around the Lothians and Fife to provide its food supply. Thirdly, the chief market for the re-exported tobacco was the Continent and the industry looked for ways of carrying its produce to that market more easily and cheaply. These three reasons combined to persuade the City of Glasgow to take the initiative in 1767 and propose a number of ventures, including the cutting of a small canal to the Forth to further trade to the advantage of the city.

Edinburgh, too, was clearly concerned with economic factors when the city challenged the Glasgow plan in 1768. Supported by the grain-producing burghs of the east, the capital argued that, far from Glasgow being the main user of the Canal, the chief goods that would be carried were agricultural, especially on an inter-regional basis, and that a deeper canal would be more suited to this type of trade. Advocates of the large canal assembled a mass of statistics to demonstrate that eastern interests were greater than those of Glasgow and indicated that the largest single commodity that would be transported was grain going to the west, followed by coal going to Edinburgh: a case that, in time, proved to be very accurate.[115] The role that the Canal could play in United Kingdom, European and Colonial trade was also emphasised. The clash of these economic interests was largely, but not solely, responsible for the bitter arguments that ensued between the two cities.

It was also Edinburgh that introduced the notion that matters of national pride were at stake. The criticism, sometimes vitriolic, of the narrow, commercial outlook of Glasgow was less a condemnation of economic motives, about which Edinburgh was equally vocal, but rather a plea to examine the wider aspects of the scheme. The demand that as the Capital was the 'abode of civility' where the aristocracy resided, it should take the lead in the scheme, seems to indicate that there was a belief that the Capital had the resource to undertake this new venture, as it had done in the days when the Estates of the Scots Parliament met there. Civic and national leadership in commercial affairs would give the city, and its leading citizens, the opportunity to exercise their frustrated political energies in a new direction, while at the same time opening the Canal to trade with the Colonies. However, the controversy indicates more than economic rivalry, and shows the emergence of a strong challenge from Glasgow to be regarded as the new capital of Scotland because of its economic importance. The pride and prestige of Edinburgh were at stake, and the appeal to the social and intellectual status of the Capital, 'the Wise men of the East' to whom the 'Fools of the West' should yield first place, is a clear indication of the strong emotions that were aroused. This view held sway temporarily, and Edinburgh was able to proceed with the Smeaton plan for a 'grand canal'. However, as the Canal developed, Glasgow emerged as the chief city in the country and, within the Canal Company, power moved to the west and after 1775 Edinburgh was seldom consulted on company affairs.

The involvement of landowners, including most of the leading aristocratic families, with the many lawyers and merchants in the management of the Company indicates yet another change in society that the Canal encouraged. As agricultural improvement proceeded, the income of landowners increased and they sought other ways of investing their surplus money. The Canal gave them just such an investment opportunity. Even in 1768, it was realised that dividends would not arise for at least ten years, and this undermines the argument that the investors were only concerned with making a profit. It seems more likely that they were prepared to delay profits in the knowledge that the enterprise was of benefit to all, as well as themselves, in the longer term.

The ease with which the different classes – nobles, middle classes, landowners and merchants – mixed in the formation, organisation and running of the Canal indicates a breakdown of formerly rigid class barriers. Involvement in the Company's affairs made it possible for the rising middle classes to enjoy a greater social mobility. Among the Glasgow merchants who participated in the venture were George and Adam Buchanan, Campbell of Clathick, William French, James Gordon, James Hopkirk, Henry Glassford and Patrick Colquhoun, and one of the early promoters of the 'small canal', Alexander Spiers, who had married into the Dundases of Kerse. True, the increasing wealth of the merchant or lawyer enabled him to purchase land and become a member of the landowning community and to share in such political power as that conferred, but this alone did not bring acceptability among the older-established families. The opportunity for social intercourse provided by participation in the Canal, and similar ventures that followed, was important in helping to give to all classes a common goal.

The problems faced during the construction forced the Company to experiment with new methods of organisation and raising finance. Smeaton's original estimate of £150,000 was quickly subscribed by the public and, as a joint-stock venture, was administered by the Council of the Company, which made calls of up to 10% at three-monthly, or longer, intervals to finance each stage of the operation. Not all the calls were honoured, however, and the Carron Company went so far as to try to use its shares as a lever to coerce the Company into meeting the demand for a cut to Carronshore to suit the commercial interests of Carron's ironworks and shipping company. As a result, the proprietors of the Forth and Clyde Canal sought a means to enforce payments, and began to assume more extensive powers, particularly through the courts. With the most influential members residing in London, the Company could not avoid being influenced by English Law, in particular finding it convenient to adopt methods similar to those used in the House of Commons, to which body it had recourse, from time to time, to gain an extension of the powers granted under the original Act. From 1773, Edinburgh lost its predominance in the Company, and Glasgow took over the administration for a time.

The financial crisis of the '70s accelerated change in the power structure. By 1773, it was clear that Smeaton's estimates were far too low. Relying on his experience in England, Smeaton had calculated the cost of cutting, banking, puddling and lock-building on a scale that proved inadequate to overcome the geological problems that presented themselves in Central Scotland. Other delays only served to aggravate the financial difficulties. Dullatur Bog took far more time and materials than any comparable section of the Canal, and the discovery of a number of bodies in the marshes halted work for some time. Roman remains abound in the area, and when the Canal was driven through the Antonine Wall and altars were unearthed at Auchendavy, work was interrupted to allow archaeologists to examine and remove the finds. These examples of concern for heritage overcoming commercial interest typify the argument put forward by Edinburgh that matters of national pride were important, although it did not stop the destruction of part of the Roman Wall. Unfortunately, the delays also caused the building work to fall further behind schedule and increased the mounting debt. When funds were exhausted and an overdraft had accumulated, further work was not possible and the Canal Company ceased to operate, apart from essential maintenance work on the operational section that had been opened as far as Kirkintilloch. Only a separate subscription by merchants from Glasgow allowed the extension of the cut to Hamiltonhill to bring the terminus closer to the city. Until the Government changed its mind and made funds available, no progress could be made.

When the Government eventually abandoned its earlier

policy of non-involvement by making £50,000 available from the Forfeited Estates, it insisted on having some control over the Company, and, in time, decision-making passed out of the hands of the Glasgow Committee into those of the Council and Governor in London. English and parliamentary methods came to influence the organisation of the Company, while Westminster was not slow to adopt some of the administrative methods that the Company had pioneered. The need to make three copies of everything in order to allow Edinburgh, Glasgow and London to be kept informed about Company affairs led to the practice of keeping records in triplicate which became common in business. Patrick Colquhoun, to handle dealing with large-scale financial operations, introduced systems that were copied by other businesses and institutions, and which were incorporated into the Acts granting charters to the banks in 1844.

The scale of the undertaking, and the complexities of construction, together with the protracted time in building, helped to produce a new occupation in Scotland. Civil engineering grew in status during the century as the demand for experts to supervise major building projects increased. Smeaton, Brindley, Golborne and Whitworth were Englishmen, already acknowledged as leading figures in canal construction, who played leading roles in the making of the Forth and Clyde Canal. During the building, a number of Scots engineers were employed and came to the attention of the public, particularly Robert Mackell and John Laurie, and joined the already well-known James Watt as leading members of the profession. Their successors were kept fully employed with the onset of 'canal mania' in the north, their skills being demonstrated on projects like the Union and Crinan Canals, and on the road- and bridge-building that continued into the next century. Some went to the Continent where their skills were appreciated, and the legend of the 'Scots Engineer' grew. The need for adaptability in the face of early problems was passed on to the next generation, who quickly adjusted to the demands made by the railway revolution.

As Scotland's economic revolution progressed, there was increasing pressure on labour to leave the land and take up full-time, non-agricultural work. Among the earliest such were the navvies, who first appeared with the building of the canals. In the early years of construction, the Company had relied on local labour, but found that the men tended to return home to bring in the harvest, leaving construction work at a standstill for weeks. To overcome this problem, non-local labour, Highland at first and later Irish, was hired on a casual basis, usually being paid for each day's work and, it was thought, less likely to go home until the work had been completed. However, the plan was frustrated as even these groups saw harvest-time as an opportunity to vary the monotony of canal-building with some familiar farm work, and there was always a demand for day labourers at that time of year. By 1790 the Company had come to realise that the most effective way of operating was to hire full-time navvies, and there was the benefit that the construction had already given experience of this type of labour to a large number of men who would require little in the way of training.

The Forth and Clyde Canal typified Scotland in the 'Age of Improvement'. It formed part of the economic revolution by helping to create the infrastructure that was essential to allow the changes in agriculture and industry to develop their full potential. Without the transfer of grain to the growing industrial cities and industrial products to the countryside as a whole, the benefits of the improvements would not have spread throughout the country. The clash between Glasgow and Edinburgh represented a clash between the new, commercial interests and the old, parliamentary and aristocratic world of pre-Union Scotland. Its resolution was an amalgam of 'Enlightened' ideas and hard-headed commercial acumen, encouraging cooperation in projects to improve the wellbeing of the whole country. By opening up regional communications, the Canal gave local people a wider perspective and helped remove their narrow, parochial outlook, a move encouraged by those who wanted Scots to regard themselves as 'North British' and lay aside old, outdated sentiments of nationality and loyalty to the House of Stuart.

While it is true that the Forth and Clyde was never to have the military role that had been proposed in the seventeenth century, it was realised during the French Wars of the late eighteenth and early nineteenth centuries that it could be used to move troops to meet the threat of a French invasion. The waterways had been used effectively to move reinforcements from England to deal with the French invasion of the West of Ireland in 1798.[116]

By adopting *ad hoc* solutions to new administrative and financial problems, the Canal Company encouraged the development of business practices that were taken up as the most effective and efficient method of running a large-scale undertaking. It played a part also in persuading the Government to move away from its rigid policy of not intervening in commercial and industrial matters.

The making of the Canal marks a watershed in the history of Scotland. It stands between the basically rural economy with its small, domestic industry and limited trade of the early eighteenth century and the large-scale industrial-commercial society with worldwide trade that was characteristic of the end of the century. As the first undertaking on a great scale, it helped show the way ahead.

The Forth and Clyde Canal marked a major advance in transport in Scotland by allowing the movement of bulk cargoes that were not suited to transportation by road. It was quickly realised that it also provided a comfortable and reasonably cheap means of passenger transport, and this sector expanded rapidly to meet demand. While it is true that the carriage of grain, timber and coal was the staple of canal traffic, the provision of a passenger service provided a profitable diversification for the

Canal Company. The appearance of the 'Swifts' and their night versions, the 'Hoolets', raised the profile of the waterway, and they were to become its symbols in the eyes of the public. By linking with the coach network an integrated transport system was created that enabled passengers to travel throughout much of Scotland with the minimum of inconvenience.

It had been anticipated that the waterway would create employment opportunities in the area through which it passed, but its success was far in excess of what had been imagined. Initially it had been expected that the need for lock-keepers and stable hands would provide employment, while the farms would benefit from the ready movement of their produce to market in the large towns. However, others were quick to realise the benefits of the navigation, and from Grangemouth to Bowling heavy industry made use of the ready supply of water and the ease with which raw materials could be brought to the workplace and the manufactured goods taken to market. Falkirk, Kirkintilloch and Glasgow were the main beneficiaries, and their populations expanded rapidly as workers moved to take up employment in the burgeoning industries. Populations doubled, and in the case of Glasgow trebled by 1831.

Iron foundries, distilleries and shipyards became major employers, the last providing the vessels that were needed to ply the Canal. As the challenge from the railways emerged in the second third of the century, the Canal Company turned to the engineering works to provide steam-powered vessels, and these were introduced for passenger transport. Fears that the wash might damage the lining of the canal meant that there was a reluctance to make full use of the new means of propulsion, and wooden vessels remained the favoured vessels for transporting bulk cargoes. It was this failure to adapt that rendered the canals unable to compete with the railways.

After 1830, the threat from the railway companies grew at an increasing pace, despite the efforts of the canal proprietors to thwart the newcomers. Lower fares and more frequent vessels were not enough to keep passengers from the cheaper and faster trains, and even the merchants, who had been the chief instigators of the Canal originally, now saw that their interests were better served by the railways. The Forth and Clyde Canal lobby finally lost out to 'railway mania' in 1838, when the House of Commons passed the Edinburgh and Glasgow Railway Act, which authorised the building of a line to compete directly with the navigation. Although the Company responded by trying to obstruct the construction of the railway, it could not withstand the tide of popular opinion.

In a number of ways the Canal had contributed to its own demise. By demonstrating that investment in large-scale transport developments brought economic benefit not only to the users of the system, but to the wider community through which it passed, the Canal provided an argument for those who advocated building a rail network throughout the country. Unlike a canal, the railways could cover most of the country and bring economic regeneration to places that the waterways could not reach. Further, the canals had produced a generation of workers with the very skills that the railway companies needed. The civil engineers who had learned their trade by overcoming the obstacles to making the Canal could find employment making the tunnels and bridges required to carry the trains across the countryside. In the now redundant labourers, the companies had a ready-made and skilled workforce of navvies to create the new system.

When the Edinburgh and Glasgow and Central Railways linked at Greenhill, just east of Croy and north of the Forth and Clyde Canal, in 1848, they opened up a passenger and goods network that stretched as far as Aberdeen in the north and London in the south. Passengers were the first to utilise the new route, and passenger travel on the Canal ended almost at once, leaving only goods carriage on the waterway. The advantage of cheap and rapid transport throughout the United Kingdom, however, was attractive to many of the Canal's customers who switched to rail wagons from the barges, although the Canal remained profitable for a further decade.

In 1849 the newly renamed Forth and Clyde Navigation merged with the Edinburgh and Glasgow Railway, and immediately control passed to the railway interest. It was feared that the new company would allow the waterway to fall into disrepair and use this as an excuse to close it. To ensure the future of the canals, the Government passed the Railway Act of 1873, which made it mandatory for the railway companies to properly maintain their canals. This Act kept the Forth and Clyde Canal in operation until the start of the twentieth century, and fishing vessels continued to use it to pass from the fishing grounds in the west to those in the east, while pleasure boats operated on the western stretch as far as Craigmarloch.

During the First World War the sea locks at Grangemouth were closed, preventing through passage to the Forth, and signalling the intention of the railway companies to put their rivals out of business. The closure of the link with the Edinburgh Union Canal in 1923 marked the end of travel directly to the Capital by water, although the Canal was not filled in as the businesses along its route required the water supply. The Forth and Clyde now struggled for survival as income fell steadily to levels far below that required to maintain it. As a consequence, many of the structures, principally the wooden lock gates, began to deteriorate.

Nationalisation of the transport system by the post-war Labour Government led to the canals being transferred to the authority of the British Transport Commission. In the debate over the relative merits of rail and road transport, dominated by the latter, little or no attention was paid to the waterways, except by those who saw them as potential motorway routes for the expanding road haulage industry. This group used the incidence of children being drowned in the neglected canals as propaganda to further their design, under the guise of removing a public danger. By 1960 they had succeeded sufficiently for the Monkland Canal to be filled in east of Cut of Junction and converted into a section of the M8 motorway.

The final straw for the Forth and Clyde was the publication of a report that claimed that the Scottish canals were no longer viable as commercial enterprises. Thereafter, the Forth and Clyde was allowed to fall into disrepair, despite the valiant efforts of the British Waterways Board to maintain it. Neglected by the public, the navigation became forgotten, even by people living in proximity to it. Yet anglers continued to fish along its banks, walkers found its towpath a peaceful retreat and the owners of small boats made use of its waters, as did some who chose it to dispose of their rubbish.

A number of factors led to a reawakening of interest after the 1960s. The growing interest in Scotland's industrial past led to study of the Canal and its history, and by 1971 the creation of the Scottish Inland Waterways Association led to a campaign to restore it, at least partially. Strathclyde Regional Council was formed in 1975 and set about making use of the Canal for recreational purposes. Working with British Waterways, the Council and the newly formed Forth and Clyde Canal Society gave the old Canal a new public image and encouraged the use of the navigation by vessels made available by the Clyde Port Authority. Volunteers began to clean up and repair the Canal, and their efforts resulted in the reopening of the stretch between Temple and Kirkintilloch in 1988. When Glasgow became the Cultural Capital of Europe in 1990, the Canal achieved its highest profile yet, and attracted thousands of visitors. That year, to celebrate the 200th anniversary of the opening of the navigation, canal boats carried water from the Clyde to empty into the Forth, re-enacting the ceremony of 1790.

The success of these events led to ambitious proposals to re-open the entire length of the Canal to navigation. Despite the enormous problems of infilled sections, damaged locks and low road bridges, the scheme attracted financial support from a number of organisations, in particular the European Commission and the National Lottery, which enabled work to start in 1999. It was decided to restore the link with the Edinburgh Union Canal at Falkirk by means of the unique 'Wheel' that would be a visitor attraction, and work commenced there also.

In May 2001 the Canal was reopened and a flotilla of craft passed from Grangemouth to Bowling in a colourful cavalcade. Two months later, Scottish Enterprise and West Dunbartonshire Council announced plans to redevelop Clydebank, using the Forth and Clyde Canal as a central feature of the scheme, and in so doing fulfilled one of the objectives of the Millennium Link, that of improving the environment.

The formal opening of the visitor centre and the Falkirk Wheel to restore the link between the Forth and Clyde and Edinburgh Union Canals by Her Majesty the Queen on 24 May 2002 marked the completion of a mammoth undertaking. Perhaps it was fitting that on her Golden Jubilee she extended royal approval to the old navigation that her predecessor, George III, authorised and which became a major landmark in the commercial, industrial and transport history of Scotland.

The reopening of the Canal has offered greater scope for trips on the boats of the Seagull Trust. This organisation allows access to canals for handicapped persons, and has been supported by shipbuilders – two of the vessels are named after the apprentices who designed and built the boats at the old Fairfield Yard in Govan and at Yarrow's – the *Govan Seagull* (based at Falkirk) and the *Yarrow Seagull* (based at Kirkintilloch). The Trust offers the opportunity for the disabled to enjoy cruises and picnics in the Campsies and to travel on the 'Wheel': in 2003, over 2500 disabled passengers were carried, and there are vessels fitted out to allow overnight voyages. With their able-bodied friends, the handicapped have contributed to the enormous increase in traffic on the canals.

With increasing numbers of vessels now using the waterway, a need has arisen for more extensive berthing facilities in Glasgow. Glasgow District Council has responded to this by heading a project team to extend the terminus from Spiers Wharf to the Pinkston Basin at Port Dundas, almost where the Forth and Clyde met the Monkland Canal. The £5-million scheme will be financed by the District Council, with £2.7 million from the city's growth fund announced by the Scottish Executive, and with a further £2.3 million being provided by the European Regional Development Fund. It is expected that this will lead to a major housing, business and leisure development of the area that will cost some £80 million over the next 10 to 15 years.[117]

The new millennium has given a fresh lease of life to the Forth and Clyde Canal, and reawakened interest in it and its associated waterways. It has also, significantly, restarted a programme of development along the reaches of the navigation that brings to fulfilment one of the chief aims of those who proposed the canal over two centuries earlier. As more use is made of the improved facilities, it is not unreasonable to suggest that the ghosts of the past must be contented that their plans are being realised, and the people of today must be grateful for the foresight of their ancestors. It now remains for the people of Scotland to make use of it as was intended.

APPENDIX A

LIST OF PROPRIETORS OF THE FORTH & CLYDE NAVIGATION

As they stand In the Act of Parliament, with the sums for which they have subscribed annexed to their names respectively.

List of Subscribers

John, Duke of Bedford	£1,000
George, Duke of Marlborough	1,000
Henry, Duke of Buccleuch	2,000
Alexander, Duke of Gordon	2,000
Charles, Duke of Queensberry and Dover	2,000
John, Duke of Roxburgh	1,000
William, Marquis of Lothian	1,000
John, Marquis of Lorne	1,000
James, Earl of Errol	1,000
James, Earl of Morton	1,000
James, Earl of Abercorn	1,000
Thomas, Earl of Kinnoul	500
Charles, Earl of Elgin and Kincardine	1,000
James, Earl of Findlater and Seafield	2,000
George, Earl of Aberdeen	1,000
William, Earl of Ruglen and March	1,000
Hugh, Earl of Marchmont	3,000
Neil, Earl of Rosebery	1,000
John, Earl of Glasgow	500
John, Earl of Bute	2,000
John, Earl of Hopetoun	2,000
Patrick, Lord Elibank	1,000
William, Earl of Panmure of the Kingdom of Ireland	1,000
James, Earl of Fife of the Kingdom of Ireland	1,000
Robert, Earl of Catherlough of the Kingdom of Ireland	2,000
Rt. Hon. Frederick Campbell (commonly called Lord Frederick)	500
Rt. Hon. James Hope (called Lord Hope)	1,000
Rt. Hon. James Stuart Mackenzie (Lord Privy Seal of Scotland)	1,000

List of Subscribers (cont.)

Rt. Hon. George Grenville	1,000
Rt. Hon. James Oswald	1,000
Rt. Hon. James Montgomery (Lord Advocate of Scotland)	500
Rt. Hon. Gilbert Laurie (Lord Provost of the City of Edinburgh)	500
George Murdoch, Provost of the City of Glasgow	1,000
Hon. James Wemyss	1,000
Hon. Col. Alexander Mackay	500
Hon. Alexander Murray	1,000
Hon. Charles Elphinstone	100
Sir John Anstruther	1,000
Sir Alexander Gilmour	1,000
Sir Alexander Ramsay	500
Sir James Dunbar	500
Sir Robert Dalziel	500
Sir George Suttie	500
Sir Alexander Grant	500
Sir William Murray	1,000
Sir James Clark	500
Sir John Hall	1,000
Sir Lawrence Dundas	10,000
Sir William Mayne – Baronets	500
Sir Fletcher Norton	1,000
Sir James Douglas	500
Sir John Lindsay – Knights	500
Admiral Francis Holbourne	1,000
Maj. Gen. David Graeme	1,000
Col. John Scott	1,000
Col. James Masterton	1,000
Col. Hector Monro	1,000
Pryce Campbell	500
Archibald Edmonston	500
Thomas Dundas of Fingask	1,000
James Grant	500
William Pulteney	1,000
Thomas Dundas the younger of Kerse	1,000
Adam Drummond	1,000
James Abercrombie	1,000
James Couts	2,000

Alexander Forrester	500
John Ross Mackye	1,000
Norman Macleod	1,000
James Veitch of Elliock	1,000
Robert Bruce of Kennet	500
John Drummond	1,000
Richard Oswald	1,000
John Mill	1,000
James Murray	1,000
Alexander Wedderburn	500
Luke Scrafton	1,000
John Purling	1,000
Hugh Ross	1,000
Joseph Salvadore	1,000
John Steuart	1,000
Robert Udney	500
John Johnstone	1,000
John Farquhar Kinloch	500
John Calcroft	4,000
Patrick Millar	500
John Cockburn	500
William Neill	1,000
Andrew Moffat	500
Alexander Spiers	1,000
William Manson	500
Henry Isaac	1,000
James Baird	500
George Glasgow & others	4,000
Andrew Clarke	1,000
Hugh Seton	1,000
James Garbet	500
David Barcley, Junr.	500
William Elliot	500
George Inglis	500
Alexander Garden	500
George Campbell	500
Alexander Kincaid	500
John Mackenzie	500
Henry Dundas	500

List of Subscribers (cont.)

James Stewart	500
John Inglis	500
John Pringle	500
Archibald Stirling	1,000
Thomas Coutts	500
John Clerk	500
George Chalmers	1,000
Andrew Millar	500
Thomas Halsey	1,000
James Hamilton	500
Thomas Lachray	500
Gibb Crawford	1,000
William Geddes	1,000
George Ross	1,000
Alexander Hunter	500
Major Chalmers	500
? James Mullen	500
John Fordyce	500
Rt. Hon. Lord Provost of Edinburgh for the City	2,000
Alex. Spiers, in trust for different people	5,000
John Pringle Esq. in trust for	
£500 to Sir Arch. Grant,	
£500 to Jas. Jackson,	
£500 to David Loch,	
£100 to Dr. Hope	1,600
	£128,700
Sir Laurence Dundas, in trust for the whole Coy. of	
Props. per Minute of G.M. 14/3/1768	21,300
	£150,000

(Source: Act of 8 George III, c. 65, 1768 in House of Lords Record Office)

APPENDIX B

TABLE OF CALLS
AND GLASGOW PAYMENTS

Minutes of the Canal Company			Glasgow Agent's Payment Records		
Date	Amount	Due By	Date	Call No.	Payment
27 May 1767	5%	25 Dec 1767			
11 Dec 1767	5%	1 Feb 1768	17 June 1768	1 & 2	£100
			27 Aug 1769	3	50
			23 Dec 1769	4	50
2 Feb 1770	5%	15 May 1770	2 Aug 1770	5	50
11 Jun 1770	10%	?	12 Sept 1770	6	100
28 Nov 1770	?	'as advertised'	no date	7	100
17 Apr 1771	?	'as advertised'	no date (1771)	8	100
			28 Dec 1772	9	100
23 Apr 1773	?	'in arrears'	10 Feb 1773	10	100
			10 May 1773	11 & 12	200
			no date (1778)	13	50

(Source: D.A.R. Forrester, *The Great Canal*, p. 13)

From the above, it seems that the first five calls were for 5% and calls 6 to 12 were for 10%, with the final one for 5%.

Glasgow appears to have been late with almost all the calls, but the lack of 'Due By' dates makes it difficult to assess how late payments were, although it is clear that the first two calls were six and four months late, and the last was not made until after the due date and the account was in arrears. Since the City was one of the better payers, the table does demonstrate one of the reasons why the Canal Company was always chronically short of funds.

APPENDIX C

SOURCES OF INCOME

YEAR	PASSENGERS	GRAIN	COAL	TIMBER
1802	–	£ 3,471	£ 1,801	£ 4,619
1805	–	3,841	3,437	4,267
1808	£ 580	6,022	4,247	1,437
1810	2,449	7,163	4,472	4,947
1812	3,453	4,674	4,588	1,986
1814	6,182	10,903	3,755	3,388
1818	7,819	13,220	4,440	4,037
1820	7,810	8,287	2,680	1,417
1822	7,933	12,028	3,475	2,828
1824	5,696	12,261	3,514	5,105
1826	4,217	11,466	3,643	3,048
1828	5,999	11,700	1,713	2,852

J Lindsay, *Canals of Scotland*, p. 219

REVENUE AND EXPENDITURE

Year	Revenue	Expenditure	Year	Revenue	Expenditure
1802	£ 23,371	£ 9,422	1878	£ 57,128	
1806	26,955	11,877	1884	56,049	£ 21,279
1820	37,215	16,178	1888	44,038	
1825	45,936	13,169	1893/4	42,601**	19,259**
1830	53,881	16,334	1898	38,281	
1834	51,167	13,181	1904	41,155	22,609
1846	54,118*	18,068*	1906	40,108	
1849	54,770*	24,555*	1913	35,136	
1855	51,805*	15,771*	1925	13,434	
1860	52,744*	14,996*	1938	16,645	23,849
1866	84,451		1951	27,838	
1868	87,145		1955	29,545	65,286
1876	65,309		1956	36,456	99,291

Source: J Lindsay, *Canals of Scotland*, pp. 220–223 (adapted)

Notes: *Figures for April to September only.

**Composite figures for 1893 and 1894.

MILLENNIUM PARTNERS

LOCAL AUTHORITIES:
  City of Edinburgh Council
  East Dunbartonshire Council
  Falkirk Council
  Glasgow City Council
  Norh Lanarkshire Council
  West Dunbartonshire Council
  West Lothian Council

PUBLIC SECTOR:
  British Waterways
  Scottish Canals
  Central Scotland Trust
  East of Scotland European Partnership
  Millennium Commission
  Scottish Natural Heritage
  Strathclyde European Partnership

ENTERPRISE COMPANIES:
  Scottish Enterprise
  Scottish Enterprise Dunbartonshire
  Scottish Enterprise Edinburgh and Lothians
  Scottish Enterprise Forth Valley
  Scottish Enterprise Glasgow
  Scottish Enterprise Lanarkshire

VOLUNTARY SECTOR:
  Edinburgh Canal Society
  Falkirk and District Canal Society
  Paisley Canal and Waterways Society
  Ratho Union Canal Association
  Scottish Inland Waterways Association

BOAT HIRERS:
The Bridge Inn, Ratho-Edinburgh Canal Centre
The Caledonian
Forth and Clyde Canal Society
Linlithgow Canal Centre
Lowland Narrowboats Ltd

SEA:
Seagull Trust
Thistle Hire Boats

OTHERS:
Canal and Riverboat Magazine
Sail Scotland Ltd

Source: British Waterways, *www.millennium*link.org.uk/links

NOTES

1 There are a number of books on the agricultural changes in Scotland during the seventeenth and eighteenth centuries; the following were the ones chiefly used in the present study:

I Whyte, *Agriculture and Society in Seventeenth-Century Scotland* (Edinburgh, 1979)

T C Smout, *A History of the Scottish People, 1560–1830* (Glasgow, 1969)

B Lenman, *Integration, Enlightenment and Industrialisation* (London, 1981)

R H Campbell, 'Stair's Scotland: the Social and Economic Background', in *Juridical Review* (1981) Part 2.

2 For discussion of the economic factors influencing the building of canals in general, and the Forth & Clyde in particular, see:

T M Devine, 'The Merchant Community', in R H Campbell & A S Skinner' (eds.), *Origins and Nature of the Scottish Enlightenment* (Edinburgh, 1982)

T M Devine, 'Colonial Commerce and the Scottish Economy', in *Comparative Studies of Scottish and Irish Economic and Social History, 1600–1900* (Edinburgh, no date)

G Donaldson, 'Stair's Scotland: the Intellectual Inheritance', in *Juridical Review* (1981), Part 2

N T Phillipson, 'Lawyers, Landowners and Civic Leadership in Post-Union Scotland', in *Juridical Review* (1976), Part 2

3 The role of societies and the political motives behind canal building are dealt with in:

Articles by Campbell, Donaldson and Phillipson in *Juridical Review* (above)

R H Campbell, *Scotland Since 1707* (Oxford, 1965)

A S Skinner, Introduction to Campbell & Skinner, *Origins and Nature of the Scottish Enlightenment* (Edinburgh, 1982)

N T Phillipson, 'The Scottish Enlightenment', in R Porter & M Teich, *Enlightenment in National Context* (Cambridge, 1981)

4 T C Smout, 'Scottish Landowners and Economic Growth 1650–1830', in *Scottish Journal of Political Economy*, 11 (1964), pp. 219–220

5 T M Devine, 'Colonial Commerce', in *Comparative Studies*, p. 179
6 National Archives of Scotland:
  Minutes of the Bathgate Trust, Shairp of Houston MSS, Ref. GD30/2161
  Papers *re* Midlothian Tolls, Richardson MSS, Ref. GD1/138/2
  Minutes of the Trustees for Edinburgh Roads, Ref. CO 2 4/1
7 S Smiles, *Lives of the Engineers*, vol. ii (Newton Abbot, 1969), p. 101
8 A W Ward *et al* (eds.), *Cambridge Modern History*, vol. v, *Ascendancy of France* (Cambridge, 1934), p. 14
9 J Priestley, *Navigable Rivers and Canals* (1831, reprinted Newton Abbot (1969), p. 266
10 *Ibid.*, pp. 266–73
11 J Cleland, *Annals of Glasgow* (Glasgow, 1816), pp. 298–309
12 J Lindsay, 'Promotion of the Forth and Clyde Canal: Glasgow versus Edinburgh', in *Transport History*, vol. xi (1980), pp. 3–12
13 D Martin, *Forth and Clyde Canal: A Kirkintilloch View* (Kirkintilloch, 1971), pp. 3–9
14 J K Allan, *There is a Canaal* (Falkirk, no date), pp. 3–20
15 Devine, 'Colonial Commerce', p. 179
16 Smiles, *Lives of the Engineers*, p. 92
17 C Hadfield, *The Canal Age* (Newton Abbot, 1968), p. 45
18 Allan, *There is a Canaal*, pp. 3–5
19 Journal of the House of Commons, vol. 31, 23 January 1767, p. 64
20 Extracts from the Records of the Burgh of Glasgow, vol. vii, 1767 (Glasgow, 1911), p. 237
21 Cleland, *Annals of Glasgow*, p. 298
22 J R Ward, *Finance of Canal Building in Eighteenth Century England* (London, 1974), pp. 133 and 183
23 Devine, 'Colonial Commerce', p. 179
24 Journal of the House of Commons, vol. 31, pp. 255–6
25 R H Campbell, *Carron Company* (Edinburgh, 1961), pp. 118–9
26 Extracts from the Records, vol. vii, p. 243
27 Journal of the House of Commons, vol. 31, 11 & 16 March 1767, pp. 211 & 215
28 Report of John Smeaton, 11 April 1767, in Falkirk Museum, Ref. a 96 .01. Part 1
29 Journal of the House of Commons, vol. 31, 28 April 1767, p. 327
30 Smeaton's Report, Falkirk, Ref. a 96. 01, Part IV

31 Forth and Clyde Canal MSS, Delvine Papers in the National Library of Scotland, Edinburgh, Ref. MS 1497, p. 122

32 Extracts from the Records, vol. vii (1767), pp. 243–4, 259–260 & 264–6

33 Journal of the House of Commons, vol. 31, 14 May 1767, p. 360

34 Preamble to the Act 8, George III, c. 65, House of Lords Record Office, pp. 928–9

35 Extract of Dispositions 1771–79, Falkirk Museum, Ref. a 53 .61

36 *Glasgow Journal*, 17–24 March 1768

37 Brindley, Yeoman and Golborne Report, 30 September 1768, Falkirk Museum, Ref. a 96 .01, Part IV

38 Smeaton's Report (1768), Ref. a 96 .01, Part III

39 J Lindsay, 'Robert Mackell and the Forth and Clyde Canal', in *Transport History*, iii, No. 3 (1968), pp. 289–290

40 J E Handley, *The Navvy in Scotland* (Cork, 1970), p. 46

41 Local History of the Forth and Clyde Canal (Strathclyde Regional Council/British Water Board), L & R 1177007/WB 700

42 Cleland, *Annals of Glasgow*, pp. 303–4

43 H B Miller, *History of Cumbernauld and Kilsyth* (Cumbernauld, 1980), p. 32

44 Whitworth's Report, 2nd August 1785, Falkirk Museum, Ref. a 10 .167, Part I

45 J Lindsay, *Canals of Scotland* (Newton Abbot, 1968), p. 31

46 Lindsay, 'Mackell', p. 291

47 Priestley, *Navigable Rivers*, p. 27

48 Allan, *There is a Canaal*, p. 14

49 Handley, *Navvy in Scotland*, pp. 47 & 50–1

50 *Ibid.*, p. 43

51 D A R Forrester, 'The Great Canal that Linked Edinburgh, Glasgow and London', in *Issues in Accountability*, No. 3 (Strathclyde, 1980), p. 7

52 Ecology of the Forth and Clyde Canal (S.R.C. & W.B.), L & R 1177007/WB 700

53 Martin, *Kirkintilloch View*, p. 4

54 John Anderson, *Of the Roman Wall between the Foth and Clyde: and of some of the Discoveries which have been lately made upon it.* Strathclyde University, Andersonian Library, Rare Books Department

55 G Hutton, *A Forth and Clyde Canalbum* (Ochiltree, 1995), p. 49

56 Forrester, *The Great Canal*, pp. 2–3

57 H G Graham, *The Social Life of Scotland in the Eighteenth Century* (London, 1969), p. 136

58 Forrester, *The Great Canal*, pp. 3–4, 10 & 16

59 Lindsay, *Canals of Scotland*, pp. 23–4

60 Handley, *Navvy in Scotland*, p. 47

61 Lindsay, 'Mackell', pp. 286–9 & 291–2

62 Preamble to the Act of 1768 (House of Lords Record Office), p. 933 and Smeaton's Report of 1768 (Falkirk Museum)

63 Smiles, *Lives of the Engineers*, vol. ii, p. 96

64 J Prebble, *Darien Disaster* (London, 1968), pp. 314–5

65 Report of Brindley, Yeoman and Golborne, 1768 (Falkirk Museum: Ref. a 96. 01 Part IV)

66 J R Ward, *Finance of Canal Building in Eighteenth Century England* (London, 1974), p. 133

67 Extracts from the Records, vol. vii, pp. 399–400

68 Cleland, *Annals of Glasgow*, p. 300

69 Forrester, *The Great Canal*, p. 13

70 Minutes of Meeting in St Alban's Tavern, 14 March 1768, Delvine Papers in National Library of Scotland, Ref. MS 1497, p. 122

71 Forrester, *The Great Canal*, pp. 5–6 & 14

72 Letter of John Seton, 24 February 1773, Delvine Papers (NLS, MS 1497), p. 138

73 Campbell, *Carron Company*, p. 118

74 Letter of C. Gascoigne to the Proprietors of the Forth and Clyde Navigation, 14 March 1768, Delvine Papers (NLS, MS 1409), pp. 122–3

75 Campbell, *Carron Company*, pp. 121–2

76 Forrester, *The Great Canal*, p. 5

77 Extract of Dispositions, 1779 (Falkirk Museum, Ref. a 53. 61)

78 Cleland, *Annals of Glasgow*, p. 301

79 Forrester, *The Great Canal*, p. 6

80 Informations for Robert Lang vs. Walter Logan, to Lord Genlee (Falkirk Museum, Ref. a 10. 167)

81 Lindsay, *Canals of Scotland*, pp. 24–5

82 Forrester, *The Great Canal*, p. 5 and Handley, *Navvy in Scotland*, p. 36

83 Allan, *There is a Canaal*, p. 14

84 Forrester, *The Great Canal*, pp. 5 & 15

85 Handley, *The Navvy in Scotland*, p. 47

86 Allan, *There is a Canaal*, p. 14

87 Notes on General Meeting of 3 May 1768, Delvine Papers (NLS), pp. 123 & 131–4

88 Forrester, *The Great Canal*, pp. 8 & 21

89 Lindsay, 'Mackell', p. 285
90 Allan, *There is a Canaal*, p. 8
91 Lindsay, *Canals of Scotland*, p. 31 and Cleland, *Annals of Glasgow*, p. 300
92 Forrester, *The Great Canal*, pp. 14–5
93 *Ibid.*, p. 12
94 Campbell, *Scotland Since 1707*, pp. 50 & 89
95 Forrester, *The Great Canal*, p. 17
96 *Ibid.*, p. 18
97 A M Smith, 'State Aid to Industry: an Eighteenth Century Example', in T M Devine (ed.), *Lairds and Improvement in Scotland in the Enlightenment* (Glasgow, 1978), p. 54
98 Forrester, *The Great Canal*, p. 16
99 G. Hutton, *Monkland Canal* (Ochiltree, 1990)
100 Alison Massey, *The Edinburgh and Glasgow Union Canal* (Falkirk Museum, 1983), pp. 6–7
101 Paul Carter, *The Forth and Clyde Canal Guidebook* (Glasgow, 1991), p. 37
102 *Ibid.*, p. 42
103 *New Statistical Account of Scotland*, County of Dumbarton, Parish of Kirkintilloch, p. 189
104 Hutton, *Monkland Canal*, p. 25
105 Don Martin & A A Maclean, *Edinburgh and Glasgow Railway Guidebook* (Strathkelvin District Library, 1992), pp. 31–2
106 *Ibid.*, p. 8
107 G Hutton, *Forth & Clyde Canalbum*, p. 29
108 Martin & Maclean, *Edinburgh & Glasgow Railway*, p. 49
109 J Lindsay, *Canals of Scotland*, pp. 210–211
110 Memories, Strathclyde Regional Council, Leisure and Recreation Department 1177007/WB8700
111 David Bolton, *Race Against Time* (London, 1990), p. 190
112 P. Carter, *Forth & Clyde Guide Book*, p. 58
113 G Hutton, *Forth & Clyde Canalbum*, p. 40
114 *Evening Times*, 12 July 2001
115 T M Devine, *The Scottish Nation, 1700–2000* (London, 2000), p. 141
116 T J Dowds, *The French Invasion of Ireland, 1798* (Dublin, 2000), p. 34
117 *The Herald*, 10 September 2003

FURTHER READING

1. The following publications deal specifically with the Forth and Clyde Canal and give details of the building and subsequent history of the canal:

Allan, J K, *There is a Canaal* (Falkirk, no date)

Martin, D, *The Forth and Clyde Canal: A Kirkintilloch View* (Kirkintilloch, 1971)

Forth and Clyde Canal Society, *The Forth and Clyde Canal Guidebook* (Glasgow, 1991)

Forrester D A R, 'The Great Canal that Linked Edinburgh, Glasgow and London', in *Issues in Accountability*, No. 3 (Strathclyde, 1980)

Hutton, G, *A Forth and Clyde Canalbum* (Ochiltree, 1995) – has an excellent collection of pictures of the Canal.

Lindsay, J, 'Robert Mackell and the Forth and Clyde Canal', in *Transport History*, Vol. i, 1968

Lindsay, J, 'Promotion of the Forth and Clyde Canal', in *Transport History*, Vol. xi, 1981

2. The canal is given some mention in the following works:

Bolton, D, *Race Against Time* (London, 1990)

Campbell, R H, *Scotland Since 1707* (Oxford, 1965)

Campbell, R H, *Carron Company* (Edinburgh, 1961)

Devine, T M, *The Scottish Nation, 1700–2000* (London, 2000)

Devine, T M, 'Colonial Commerce and the Scottish Economy', in *Comparative Studies of Scottish and Irish Economic and Social History* (Edinburgh, no date)

Hadfield, C, *The Canal Age* (Newton Abbot, 1968)

Hutton, G, *Monkland: The Canal that Made Money* (Ochiltree, 1990)

Lindsay, J, *The Canals of Scotland* (Newton Abbot, 1968)

Martin, D. & Maclean, A A, *Edinburgh and Glasgow Railway Guidebook* (Strathkelvin, 1992)

Massey, A, *The Edinburgh and Glasgow Union Canal* (Falkirk, 1983)
Priestley, J, *Navigable Rivers and Canals* (originally published in 1831, reprinted Newton Abbot, 1969)
Rolt, L T C, *Navigable Waterways* (London, 1969)
Ward, J R, *Finance of Canal Building in Eighteenth Century England* (London, 1974)

3. A very useful series of booklets was produced by Strathclyde Regional Council's Leisure and Recreation Department in cooperation with the British Waterways Board, ref. L & R 1177007/WB 700: Ecology of the Forth and Clyde Canal; Geology of the Forth and Clyde Canal; Local History of the Forth and Clyde Canal.

4. Primary sources are scattered over a number of locations, the following being the ones most useful for the present work:

Falkirk Museum: holds the reports by Smeaton 1767 and 1768 (a 96 . 01, in 4 parts); Brindley, Yeoman and Golborne, 1768 (a 96 . 01); Whitworth, 1785, (a 10 . 167/1); Extracts of Dispositions, 1779 (a 56 . 61) and the Informations for Robert Lang v. Walter Logan (a 10 . 167/3)
House of Lords Record Office has details of the various Acts relating to the Forth and Clyde Canal
National Library of Scotland, Edinburgh has the Journal of the House of Commons (for all relevant years); Delvine Manuscripts (MS 1497); copies of the *Edinburgh Evening Courant* and the *Scots Magazine*, vols. xxix–liii (1767–90)
Mitchell Library, Glasgow holds copies of *Glasgow Courant* for 1767 and the *Glasgow Journal* for 1767–90; Extracts from the Records of the Burgh of Glasgow, vols. vii and viii
Andersonian Library Rare Books Department, Strathclyde University, Glasgow has John Anderson's notebook containing the papers he presented to the Glasgow Literary Society in 1770 and 1773, on the Roman finds that the Canal Company presented to the University of Glasgow
Stirling University Library holds copies of the *Scots Magazine*, vols. xxix–liii

National Library of Scotland, Edinburgh holds the Delvine Manuscripts
and copies of the *Edinburgh Evening Courant* and the Journal of the House
of Commons

National Archives of Scotland, Edinburgh holds Shairp of Houston MSS,
Richardson MSS and the Minutes of the Trustees for the Edinburgh
Roads

British Waterways Board/Strathclyde Regional Council booklets on the
Ecology, Geology and Local History of the Forth and Clyde Canal

William Patrick Library, Kirkintilloch, has an extensive archive of photo-
graphic material relating to the Canal in general and Kirkintilloch in
particular.

INDEX